About the author

Nicholas Maxwell has devoted much of his working life to arguing that we need to bring about a revolution in academia so that it comes to seek and promote wisdom and does not just acquire knowledge. Apart from the present book, he has published four others on this theme: *From Knowledge to Wisdom* (Blackwell, 1984; 2nd ed., Pentire Press, 2007), *The Comprehensibility of the Universe* (Oxford University Press, 1998), *The Human World in the Physical Universe* (Rowman and Littlefield, 2001) and *Is Science Neurotic?* (Imperial College Press, 2004). He has also contributed to a number of other books, and has published numerous papers in science and philosophy journals on problems that range from consciousness to quantum theory. A book discussing his work has been published: *Science and the Pursuit of Wisdom: Studies in the Philosophy of Nicholas Maxwell*, edited by Leemon McHenry (Ontos Verlag, 2009). For nearly thirty years Maxwell taught philosophy of science at University College London, where he is now Emeritus Reader in Philosophy of Science and Honorary Senior Research Fellow. He has given lectures at Universities and Conferences all over Britain, Europe and north America. He has taken part in the BBC Programme 'Start the Week' on Radio 4, and the Canadian Broadcasting Corporation 'Ideas' Programme 'How To Think About Science'. In 2003 he founded Friends of Wisdom, an international group of people sympathetic to the idea that academic inquiry should help humanity acquire more wisdom by rational means (see www.knowledgetowisdom.org). More information about his life and work can be found on his website: see www.nick-maxwell.demon.co.uk.

Also by the author

From Knowledge to Wisdom
The Comprehensibility of the Universe
The Human World in the Physical Universe
Is Science Neurotic?

WHAT'S WRONG

WITH SCIENCE?

Towards a People's Rational Science of
Delight and Compassion

Nicholas Maxwell

Pentire Press

Pentire Press

Pentire Press
13 Tavistock Terrace
London N19 4BZ
editor@pentirepress.plus.com

First published by Bran's Head Books 1976

Second edition 2009

A catalogue record for this book
is available from the British Library

ISBN   978-0-9552240-1-0

Printed by Lightning Source

CONTENTS

DIAGRAMS

Dedication of the 1976 Edition

My thanks to my dearest friends, for their interest, criticisms, furious objections, suggestions, and above all, help, support and encouragement:

Chris, Dave, Neash, Rich, Evadne, Brian, Nancy, Carol, Larry, Jeremy, Paul, Dot, Philip, Richard, Ingrid, Andrew – and many others.

This second edition is dedicated to members of Friends of Wisdom

PREFACE TO SECOND EDITION

This book spells out an idea that just might save the world. It is that science, properly understood, provides us with the methodological key to the salvation of humanity.

A version of this idea can be found buried in the works of Karl Popper. Famously, Popper argued that science cannot verify theories, but can only refute them. This sounds very negative, but actually it is not, for science succeeds in making such astonishing progress by subjecting its theories to sustained, ferocious attempted falsification. Every time a scientific theory is refuted by experiment or observation, scientists are forced to try to think up something better, and it is this, according to Popper, which drives science forward.

Popper went on to generalize this falsificationist conception of scientific method to form a notion of rationality, critical rationalism, applicable to all aspects of human life. Falsification becomes the more general idea of criticism. Just as scientists make progress by subjecting their theories to sustained attempted empirical falsification, so too all of us, whatever we may be doing, can best hope to achieve progress by subjecting relevant ideas to sustained, severe criticism. By subjecting our attempts at solving our problems to criticism, we give ourselves the best hope of discovering (when relevant) that our attempted solutions are inadequate or fail, and we are thus compelled to try to think up something better. By means of judicious use of criticism, in personal, social and political life, we may be able to achieve, in life, progressive success somewhat like the progressive success achieved by science. We can, in this way, in short, learn from scientific progress how to make personal and social progress in life. Science, as I have said, provides the methodological key to our salvation.

I discovered Karl Popper's work when I was a graduate student doing philosophy at Manchester University, in the early 1960s. As an undergraduate, I was appalled at the triviality, the sterility, of so-called "Oxford philosophy". This turned its back on all the immense and agonizing problems of the real world – the mysteries and grandeur of the universe, the wonder of our life on earth, the dreadful toll of human suffering – and instead busied itself with the trite activity of analysing the meaning of words. Then I discovered Popper, and breathed a sigh of relief. Here was a philosopher who, with exemplary intellectual integrity and passion, concerned himself with the profound problems of human existence, and had extraordinarily original and fruitful things to say about them. The problems that had tormented me had in essence, I felt, already been solved.

But then it dawned on me that Popper had failed to solve his fundamental problem – the problem of understanding how science makes progress. In one respect, Popper's conception of science is highly unorthodox: all scientific knowledge is conjectural; theories are falsified but cannot be verified. But in other respects, Popper's conception of science is highly orthodox. For Popper, as for most scientists and philosophers, the basic aim of science is knowledge of truth, the basic method being to assess theories with respect to evidence, nothing being accepted as a part of scientific knowledge independently of evidence. This orthodox view – which I came to call standard empiricism – is, I realised, false. Physicists only ever accept theories that are unified – theories that depict the same laws applying to the range of phenomena to which the theory applies. Endlessly many empirically more successful disunified rivals can always be concocted, but these are always ignored. This means, I realised, that science does make a big, permanent, and highly problematic assumption about the nature of the universe independently of empirical considerations and even, in a sense, in violation of empirical considerations – namely, that the universe is such that all grossly disunified theories are false. Without some such presupposition as this, the whole empirical method of science breaks down.

It occurred to me that Popper, along with most scientists and philosophers, had misidentified the basic aim of science. This is not truth per se. It is rather truth presupposed to be unified, presupposed to be explanatory or comprehensible (unified theories being explanatory). Inherent in the aim of science there is the metaphysical – that is, untestable – assumption that there is some kind of underlying unity in nature. The universe is, in some way, physically comprehensible.

But this assumption is profoundly problematic. We do not know that the universe is comprehensible. This is a conjecture. Even if it is comprehensible, almost certainly it is not comprehensible in the way science presupposes it is today. For good Popperian reasons, this metaphysical assumption must be made explicit within science and subjected to sustained criticism, as an integral part of science, in an attempt to improve it.

The outcome is a new conception of science, and a new kind of science, which I called aim-oriented empiricism. This subjects the aims, and associated methods, of science to sustained critical scrutiny, the aims and methods of science evolving with evolving knowledge. Philosophy of science (the study of the aims and methods of science) becomes an integral, vital part of science itself. And science becomes much more like natural philosophy in the time of Newton, a synthesis of science, methodology, epistemology, metaphysics and philosophy.

The aim of seeking explanatory truth is however a special case of a more general aim, that of seeking valuable truth. And this is sought in order that it be used by people to enrich their lives. In other words, in addition to metaphysical assumptions inherent in the aims of science there are value assumptions, and political assumptions, assumptions about how science should be used in life. These are, if anything, even more problematic than metaphysical assumptions. Here, too, assumptions need to be made explicit and critically assessed, as an integral part of science, in an attempt to improve them.

Released from the crippling constraints of standard empiricism, science would burst out into a wonderful new life, realising its full potential, responding fully both to our sense of wonder and to human suffering, becoming both more rigorous and of greater human value.

And then, in a flash of inspiration, I had my great idea. I could tread a path parallel to Popper's. Just as Popper had generalized falsificationism to form critical rationalism, so I could generalise my aim-oriented empiricist conception of scientific method to form an aim-oriented conception of rationality, potentially fruitfully applicable to all that we do, to all spheres of human life. But the great difference would be this. I would be starting out from a conception of science – of scientific method – that enormously improves on Popper's notion. In generalizing this, to form a general idea of progress-achieving rationality, I would be creating an idea of immense power and fruitfulness.

I knew already that the line of argument developed by Popper, from falsificationism to critical rationalism, was of profound importance for our whole culture and social order, and had far-reaching implications and application for science, art and art criticism, literature, music, academic inquiry quite generally, politics, law, morality, economics, psychoanalytic theory, evolution, education, history – for almost all aspects of human life and culture. The analogous line of argument I was developing, from aim-oriented empiricism to aim-oriented rationalism, would have even more fruitful implications and applications for all these fields, starting as it did from a much improved initial conception of the progress-achieving methods of science.

The key point is extremely simple. It is not just in science that aims are profoundly problematic. This is true in life as well. Above all, it is true of the aim of creating a good world – an aim inherently problematic for all sorts of more or less obvious reasons. It is not just in science that problematic aims are misconstrued or "repressed"; this happens all too often in life too, both at the level of individuals, and at the institutional or social level as well. We urgently need to build into our scientific institutions and activities the aims-and-methods-improving methods of aim-oriented empiricism, so that scientific aims and methods improve as our scientific knowledge and understanding improve. Likewise, and even more urgently, we need to build into all our other institutions, into the fabric of our personal and social lives, the aims-and-methods-improving methods of aim-oriented rationality, so that we may improve our personal, social and global aims and methods as we live.

One outcome of the 20th century is a widespread and deep-seated cynicism concerning the capacity of humanity to make real progress towards a genuinely civilized, good world. Utopian ideals and programmes, whether of the far left or right, that have promised heaven on earth, have led to horrors. Stalin's and Hitler's grandiose plans led to the murder of millions. Even saner, more modest, more humane and rational political programmes, based on democratic socialism, liberalism, or free markets and capitalism, seem to have failed us. Thanks largely to modern science and technology, many of us today enjoy far richer, healthier and longer lives than our grandparents or great grandparents, or those who came before. Nevertheless the modern world is confronted by grave global problems: the lethal character of modern war, the spread and threat of armaments, conventional, chemical, biological and nuclear, rapid population growth, severe poverty of millions in Africa, Asia and elsewhere, destruction of tropical rain forests and other natural habitats, rapid extinction of species, annihilation of languages and cultures. And over everything hangs the menace of climate change, threatening to intensify all the other problems (apart, perhaps, from population growth).

All these grave global problems are the almost inevitable outcome of the successful exploitation of science and technology plus the failure to build aim-oriented rationality into the fabric of our personal, social and institutional lives. Modern science and technology make modern industry and agriculture possible, which in turn make possible population growth, modern armaments and war, destruction of natural habitats and extinction of species, and global warming. Modern science and technology, in other words, make it possible for us to achieve the goals of more people, more industry and agriculture, more wealth, longer lives, more development, housing and roads, more travel, more cars and aeroplanes, more energy production and use, more and more lethal armaments (for defence only of course!). These things seem inherently desirable and, in many ways, are highly desirable. But our successes in achieving these ends also bring about global warming, war, vast inequalities across the globe, destruction of habitats and extinction of species. All our current global problems are the almost inevitable outcome of our long-term failure to put aim-oriented rationality into practice in life, so that we actively seek to discover problems associated with our long-term aims, actively explore ways in which problematic aims can be modified in less problematic directions, and at the same time develop the social, the political, economic and industrial muscle able to change what we do, how we live, so that our aims become less problematic, less destructive in both the short and long term. We have failed even to appreciate the fundamental need to improve aims and methods as the decades go by. Conventional ideas about rationality are all about means, not about ends, and are not designed to help us improve our ends as we proceed. Implementing aim-oriented rationality is essential if we are to survive in the long term. To repeat, the idea spelled out in this book, if taken seriously, just might save the world.

Einstein put his finger on what is wrong when he said "Perfection of means and confusion of goals seems, to my opinion, to characterize our age." This outcome is inevitable if we restrict rationality to means, and fail to demand that rationality – the authentic article – must quite essentially include the sustained critical scrutiny of ends.

Scientists, and academics more generally, have a heavy burden of responsibility for allowing our present impending state of crisis to develop. Putting aim-oriented rationality into practice in life can be painful, difficult and counter-intuitive. It involves calling into question some of our most cherished aspirations and ideals. We have to learn how to live in aim-oriented rationalistic ways. And here, academic inquiry ought to have taken a lead. The primary task of our schools and universities, indeed, ought to have been, over the decades, to help us learn how to improve aims and methods as we live. Not only has academia failed miserably to take up this task, or even see it as necessary or desirable. Even worse, perhaps, academia has failed itself to put aim-oriented rationality into practice. Science has met with such astonishing success because it has put something like aim-oriented empiricism into scientific practice – but this has been obscured and obstructed by the conviction of scientists that science ought to proceed in accordance with standard empiricism – with its fixed aim and fixed methods. Science has achieved success despite, and not because of, general allegiance of scientists to standard empiricism.

The pursuit of scientific knowledge dissociated from a more fundamental concern to help humanity improve aims and methods in life is, as we have seen, a recipe for disaster. This is the crisis behind all the others. It is this crisis that this book tackles head on.

Much of the book takes the form of a fierce debate between a Scientist and a Philosopher, although towards the end of the book various other characters blunder into the book – a Romantic, a Rationalist, a Liberal, a Marxist, a Christian, a Buddhist and, right at the end of the book, a Wino. I am ashamed to say that even I put in an appearance towards the end, when things get a bit out of hand.

When I wrote the book, I wanted the Scientist to be misguided, a firm upholder of the orthodox conception of science of standard empiricism, but nevertheless a man of intellectual integrity. I had in mind someone like the psychologist Hans Eysenck. My idea was that the argument should reflect real life arguments in being explosively emotional at times, and also such that no one was convinced by the arguments of the opposition. In Plato, again and again, Socrates produces ridiculous arguments and his opponents say "Yes, O Socrates" and "How true, O Socrates". In my experience this never happens in real life. My dialogue, I decided, would be the very opposite of Plato's dialogues in this respect.

I first had my "flash of inspiration", upon which this book is based, in 1972. I wrote a manuscript called *The Aims of Science* and sent it off to Macmillan's for consideration for publication. I met three editors, each of whom became very excited about the book before leaving and passing the manuscript onto their successor. Finally the book was passed onto a new editor, a Marxist I was told, who I never met, and who rejected the book. It was never published. I wrote and wrote drafts and sketches of books, one after another, in a frenzy of despair, fearing I would never succeed in publishing my great idea. Then a friend introduced me to a friend of his, who said he would publish a book of mine if I could get it ready in six weeks. I thought about it for three weeks, and then, in a state of exalted concentration, managed to write the whole of this book in the remaining three weeks. The debate between Scientist and Philosopher raged furiously in my head. I remember feeling as if I was in a train hurtling towards a dark tunnel; I had a few precious seconds to release a dove with an all-important message for humanity, but if I was not quick, the train would enter the tunnel, it would be too late, the dove would be killed, and the message would remain undelivered. I did finish the book on time, and it was published in the Autumn of 1976.

This book definitely belongs to the romantic phase of my working life. One of the accomplishments of the idea I expound is that it achieves a synthesis of rationalism and romanticism. As I say at one point: "At its best, science puts the mind in touch with the heart, and the heart in touch with the mind, so that we may acquire heartfelt minds, and mindful hearts". Nevertheless, it is difficult in practice to achieve a balance between these two wings of our culture. In subsequent work I have swung into rationalist mode, anxious to make out as cogent a case as I can for the idea I have been struggling to communicate all these years. In this book, the romantic mode prevails.

But in rereading this book for this second edition, I was delighted, but also somewhat dismayed, to discover that much of the work I thought I had propounded later, in subsequent books and articles, is already present here, even if sometimes in nascent form. What the book has to say is as relevant today as it was in 1976 – perhaps more so. Subsequent intellectual developments have not dimmed its message, and subsequent events have, if anything, only served to highlight the urgency of what it has to say. Apart from correcting typographical errors, and adding at the end a list of relevant books and articles published after 1976, I have made no changes.

We are in deep trouble. We can no longer afford to blunder blindly on our way. We must strive to peer into the future and steer a course less doomed to disaster. Humanity must learn to take intelligent and humane responsibility for the unfolding of history. I hope this book helps.

Tavistock Terrace, London, January 2009

xiv

CHAPTER ONE

A PEOPLE'S SCIENCE

Before we plunge into a discussion of what is *wrong* with science, let me begin by giving in outline a picture of what would seem to me to be an *ideal* science, a *perfect* science, a science with nothing seriously wrong with it. With such a picture of an ideal science before us, we can then go on to consider the questions: In what respects, and why, does science as it exists today fall short of the ideal? Does our picture of an ideal science really represent a true ideal for science? Is this really the direction in which we should seek to develop the science that we have at present? Is our ideal a *desirable* ideal for science? How in general ought we to go about resolving discrepancies between scientific *ideals* and scientific *practice*?

However, before I attempt to outline an ideal for science – this highly sophisticated creation of our Western civilization – I would like to begin with something that is perhaps in certain respects rather more "primitive", but which can, I believe, with justice stand for the heart, the essence, of what an ideal science ought to be. I would like to begin with the songs that the Pygmy people of central Africa sing to their forest – as described by Colin Turnbull in his book *The Forest People*.[1] These Pygmies, it seems, trust and love their world. For them, the forest is good; it cares for them, shelters them, provides them with food, with materials for huts and clothing. It is beautiful. But every now and again a tragedy happens. Someone dearly loved dies. The Pygmies believe that such tragedies occur because the forest falls asleep. And so they sing to the forest, gently to reawaken the forest. The singing takes place at night, and may go on, night after night, for two or three weeks. The singing unfolds in accordance with certain loosely observed conventions. But the kind of obsessive concern with the niceties of ritual that is found among the Bantu villagers, agricultural people who live in clearings on the edge of the forest, is wholly absent. Quite unlike the Pygmies, the Bantus distrust and fear the forest: they believe it is full of spirits, much to the amusement of the Pygmies who are wholly without superstitions of that kind. In performing their various ritualistic acts, the main concern of the Bantu villagers is to ensure that all the proper details of the ritual are enacted in their proper order, so that the relevant danger may be averted, the threat from revengeful spirits neutralized. The Pygmies' singing, however, has nothing of this kind to it whatsoever. It is not *ritual* that matters. The whole purpose of the singing is once again to put the Pygmy people into touch with that in the forest – whatever it may be – which cares for them, and which can be trusted and loved. The purpose of the singing is to reawaken within the people who participate the experience of the mystery and beauty of the forest. The singing recreates the relationship of trust and love between people and forest.

All this seems to me to constitute as good a model as any for an Ideal science. Ideally, science arises out of our endeavour to experience that which is significant and beautiful in the world around us. The purpose of science, ideally, is to establish a good relationship between people and cosmos, and people and people. Our scientific theories are our songs, designed to help us to experience that which is beautiful in Nature. And our scientific technology is, as it were, the outcome of our singing infused into Nature herself, designed to "awaken" her to our needs. Modern science and technology, at its Ideal best, is but a sophistication and elaboration of that which is practised by the Pygmy people.

It may be said that this Pygmy analogy is a false analogy just because the Pygmy theory is false: the forest does not fall asleep, and is not awakened by the Pygmies' songs. But this does not quite do justice to what the singing achieves. Something rather special does exist within those rain forests of central Africa: a community of people who love their world, and who openly share this love with one another. The Pygmies are a part of their cosmos, a part of the life of the forest: and the singing does in fact reawaken and express something very fine within that forest cosmos, namely the Pygmies' love of the forest. That exists, as much as the trees exist, and the antelope, the frogs, the rain, the slanting sunlight.

All this is as relevant to us as it is to the Pygmy people. Can we create within ourselves the kind of trusting, loving relationship to our world that the Pygmies have for their world? Can we discover in our world that which calls forth from us such a response? A proper task for an Ideal science is to reveal to us, open up for our experience, that which does deserve to be responded to in such a way.

We have moved beyond the Pygmy world, into an awareness of a cosmos with broader, more extensive horizons. And we have discovered, we believe, that Nature is in a sense utterly impersonal, blind and deaf to all our needs, our cries for help, our songs. And yet, there is a sense in which this is not quite true. We have, after all, come into existence, and flourished, tended and supported by this same blind, impersonal Nature. Even within our very scientific knowledge of Nature, there is an essential element of trust and love, an aspect of faith. In theoretical physics, for example, we invariably choose the simplest, the most harmonious, coherent, *beautiful* theory, other things being equal. But we do not *know* that Nature is mathematically beautiful in this kind of way: it rests on an act of faith, of trust. In this sense our scientific theories are our hopeful songs to Nature. And if Nature is blind and deaf to all our singing, then all the more does She need our help, our intervention, via technology, to become more sensitively "awakened" to our needs.

Thus, with some justice, we might well take up the following attitude. Our scientific theories are ideally our songs to the cosmos: and our technology, our social arrangements and institutions are ideally the outcome of the songs of our human aspirations embodied in Nature herself. Just because most of Nature is blind to our human needs and problems, in order to help her to sustain and grow to fruition that which is best in ourselves, we need to infuse into Nature herself the patterns, the music of our human aspirations. And in order to do this successfully we need to attend to the patterns of Nature herself. By subtly and delicately feeding into Nature's patterns our human patterns we help Nature to help us. Science and technology, at their ideal best, take as their aim the delicate adjustment of our relationship to Nature, so that this

3

relationship may be more harmonious, more understanding and appreciative, more humanly fruitful, less anguished and frustrated. And in order to achieve this aim we need to bring together sensitive knowledge of Nature's patterns and insight into the patterns of our most urgent needs and best human aspirations, so that we may develop a successful technology that is sensitively and delicately adjusted to promoting the best that may potentially be within us. Our scientific songs, created out of a loving attention for our world, do enable us to awaken the world to our needs. Our machines, chemicals, utensils, telephones, televisions, medicines, are parts of Nature awakened to our needs. And if these parts of Nature that are sensitively imbued with responsiveness to human need fail to please us and satisfy us, then the fault lies perhaps with us, with the fact that we have not taken our best interests sensitively and intelligently enough into account. Our very *brains* are bits of Nature sensitively adapted and responsive to the realizing of our needs, desires and aspirations. And if our brains play us up, and fail to take us to where we wish to be, then the fault may well lie with blind Nature, with those aspects of Nature that know nothing of us, and our needs; but alternatively the fault may lie with us, with the fact that we have not been able in our lives to care with sufficient perceptiveness for our brains, for ourselves. For of course we are, in part, our brains.

Is this a world of love? It is up to us! We can see ourselves as bits of Nature seeking to love herself, seeking to live with joy, delight and compassion. It is a delicate question whether we can exist joyfully with a full awareness of the world of reality. Can we love the world and each other as the world and ourselves is revealed to us in terms of our best *ideas* about the world and ourselves, in terms of our best *theories*? Is the cosmos, as revealed to us by modern science, *lovable*? Ideal science, with scrupulous honesty, objectivity, attention to detail, seeks to answer – or seeks to help us to answer – Yes!

A little more formally, my picture of an ideal science can be outlined in the following terms. Science is something that is created by people as a result of their attempts to realise desirable human ends. At its ideal best, science is the outcome of people seeking to discover, know, experience, apprehend, understand and appreciate that which is significant, interesting, fascinating, beautiful in the world around us. It is the outcome of people seeking to discover how desirable human, social aims can be realised, urgent human, personal, social problems can be solved. It is the outcome of the *sharing* of all this between people. Ideally, science is something that goes on between friends. Science has a severely *objective, impersonal* side to it; but it also has a highly personal, subjective, emotional, motivational side to it, as well: and it is important, for the ideal flowering of science at its best, that these two aspects be intimately inter-related, so that they are capable of close, intimate communication with one another. It is desirable – or rather, essential – that science shall have an objective aspect, so that science can accurately represent that which really does exist, and in a way which renders it visible, knowable and available, in principle, to anyone, and not just to a restricted few. There is, in other words, associated with science at its best, an open, democratic character, a concern to render generally accessible that which really does exist which may be of value to people, an invitation extended to anyone who may be interested to join in, and to contribute to, the common shared scientific quest. However, if this objective side of science is to serve its proper human function, it is important that it recognizes, and gives a place to, the personal, the individual, the subjective and emotional. The ultimate aim of science, we may say, is to put people into touch with the world, and into touch with each other. The task is to improve the relationship between person and cosmos, and between person and person, help make such relationships more knowledgeable, understanding, appreciative, less anguished and frustrated. At the heart of science ideally, there is the concern to help the human heart to its passionate fulfilment, in full awareness of the world of reality. The human heart *is* the heart of science. Quite clearly science can only succeed in helping to promote human happiness, personal and inter-personal fulfilment, if the *objective* aspect of science is sensitively aware of, and responsive to, the needs, desires, feelings and aspirations of people. At its best, science puts the mind in touch with the heart, and the heart in touch with the mind, so that we may acquire heartfelt minds, and mindful hearts.

5

At its best, then, science does not cater for the dissociated mind or intellect seeking merely intellectual *knowledge*. It is not created by, and for, mere *intellects*. It is the creation of people, for people, out of the desire of people to realise their aims. At the centre of science, we may say, exist people living out their lives. Science, at its best, is sensitively, delicately and intelligently arranged around people living out their lives, encouraging and promoting human life to develop in fruitful and enjoyable directions.

Objective science is as it were machinery, whether in the form of hardware, or in the form of ideas, theories, solutions to problems, that is designed to help us achieve what we really want to achieve. If it is to be of genuine value in this respect, it must clearly be designed in such a way as to take into account, sensitively and in-telligently, the needs, desires, feelings, problems, of people. A product of objective science may exist, for example, as a car, or a pill, designed to assist travel, or to help us recover from illness. Or again, a product of objective science may exist as an idea, a theory, an observation, an experiment, a problem, a solution to a problem, a technique of problem solving, designed to enhance our aware-ness, our experience, our appreciation, of significant, interesting, beautiful features of the world around us that might otherwise remain hidden from us; or designed to help us solve our problems of living, attain that which is of value, and which we desire to attain. In either case, the *value* of such a product of objective science resides in its capacity to help us realise *desirable* human desires. It will only have the capacity to be of human use and value in this way if it has been designed with a sensitive and intelligent awareness of human needs, feelings, desires and aspirations in mind. Objective science dissociated from personal subjective feel-ing, need and desire, cannot hope to serve human interests sensitively and intelligently, in this kind of way. It is for this reason that the success of science depends upon contact and communication being preserved between the objective and the subjective, the inter-personal and the personal, thought and desire, intellect and feeling, mind and heart.

On the outer fringes of science, as it were, there are scientific experts exploring and unravelling detailed, technical, intricate problems of science. The work, thought and discoveries of the experts exist however for the sake of the vast majority, the non-experts. Technical, detailed questions are delegated to experts for expert examination: the whole point of such expert examination is however to return conclusions for the assessment, understanding and use of the non-experts. Scientific experts do not own science: the centre of gravity of science lies within the community, as an aspect of its general culture, and not within the technical knowledge of experts.

In the end it is the knowledge and understanding of *people* that matters; and even more it is the quality of life of people that matters. Knowledge and understanding that exists only in technical scientific publications, without the understanding and enthusiasm of *people*, is only potential human knowledge, not actual human knowledge. The whole point of developing objective, technical scientific knowledge, recorded in an impersonal fashion, is in the end to stimulate and promote the knowledge and understanding of people, in order to enrich the lives of people. A science which had become entirely detached from people, as a result, for example, of being taken over, with immense success, by mindless robots or computers, which continued to amass an amazing range of new scientific knowledge which no person, however, could any longer understand, appreciate or experience, would not be a science meeting with amazing success at all. On the contrary, given this eventuality, science, as a human enterprise, would have come to an end. The progress of science cannot, in other words, be conceived of in purely objective, impersonal, intellectual terms; it is to be conceived of in personal, social terms. For its continued existence science needs the attention, the active concern, the love, of people. Science lives in the minds and hearts of individual people, and in the relationships between people; it does not exist on paper, magnetic tape or photographic plate. Science only comes alive with the intervention of the person.

Descartes' *Cogito ergo sum* needs to be revised much in the way once suggested by Tolstoy.[2] To misquote Descartes, and amalgamate him with Popper, we should not say: *We doubt, and therefore know*, but rather: *We desire, and therefore know*. Science is the

7

outcome, the expression, and the attempt to realise, human desire. It is at its most objective and rational best when understood, and actively pursued, in this kind of way. Science is, as it were, a profoundly optimistic and responsible expression of our faith in the value of human life, and is pursued in an attempt to enrich and enhance the value of human life. And the mystery, in the end, is perhaps just the mystery of what it is that we ultimately want, what it is that we ultimately desire.

CHAPTER TWO

REASON REQUIRES A PEOPLE'S SCIENCE

It will not have escaped the notice of anyone who has caught the drift of the remarks of the last chapter that science, as it exists today, is very different from the kind of ideal person-centred science that I have just described. By and large, science today does not seem to exist as a generally understood and appreciated part of our culture, sensitively and intelligently arranged around the needs, the problems, and the aspirations of human life. Science as we know it does not seem to be at all the direct expression of human desire, something that arises out of the concern of people to solve their problems of living, enhance and share their knowledge, understanding and appreciation of the world around us. Science does not seem to be an open, democratic, readily available part of our society and culture, accessible to all, the property of the public, as it were, as opposed to being the property of experts. It is not easy for non-scientists even to understand expert, professional science, let alone make any kind of noticed criticism of, or contribution to, such science, arrange for a non-professional human voice to be heard within the objective, intellectual domain. There does not seem to be any kind of easy interchange between the objective and the subjective, the impersonally scientific and the personal, the intellectual and the emotional, thought and desire. On the contrary, science, as it exists at present, seems to be very much dissociated from ordinary human life. The scientific, intellectual, rationalistic domain seems to be very much distinguished from, severed from, personal feelings, thoughts, desires, problems, pursuits. Science at present *does* seem to be created by and for the dissociated intellect, the dissociated mind, and not at all for the rounded, whole human being. Science, at present, seems to seek exclusively value-neutral factual knowledge and does not, in the first instance at least, seem to be concerned at all with revealing and rendering accessible to people that which is of value, helping to promote human welfare, enhancing the quality of personal life. It scarcely seems to make sense of science, as it exists at present, to say that it is sensitively and intelligently arranged around the needs, problems, desires and aspirations of people, being intelligently designed to help people realise their desirable human ends. The centre of gravity of science does not at all seem to lie with the general public; on the contrary, modern science seems to be very much the exclusive preserve of the specialist.

What has gone wrong? Why do we not have today a sensitive, intelligent person-centred science? Why is there this gulf between the ideas, problems, preoccupations, feelings, desires, values of people who are not professional scientific experts, and the ideas, problems, preoccupations and values of professional specialist science? Why is there this rift between human life and the intellectual, rationalistic domain of science?

The suggestion that I have to make in this book is that one major reason why the kind of ideal person-centred, person-oriented science that I have briefly described above has never been given a fair chance to flourish, is that the scientific community, by and large, has sought to make science conform to a bad ideal for science, a bad idea as to what it is to be scientific, a bad *philosophy* of science.

For consider how a professional scientist might reply to the suggestion that what we need is the kind of person-centred science that I have described above. He would be almost bound to declare that what I have described represents a *false* ideal for science, an appallingly irrational conception of science, an extraordinary, confused mish-mash of the intellectual and the psychological, the rational and the sociological, cultural and evaluative. "What you have described," he might argue, "scarcely represents *science* at all. It is rather common opinion, common culture, enriched perhaps with the artist's eye, the poet's sensitivity and the psychotherapist's insight. Of necessity, genuine scientific knowledge must be dissociated somewhat from common opinion, ordinary human life. For the fundamental task of science is simply to improve our knowledge and understanding of objective factual truth. As science progresses, as scientific knowledge accumulates, it is inevitable that it will become somewhat esoteric, specialized, distanced from common opinion. This may be regrettable; but it is inevitable. And indeed, to a considerable extent, the success, objectivity and authenticity of science actually *depends* upon science being dissociated somewhat from the affairs and preoccupations of ordinary human life. For the all-important condition that must be satisfied, if we are to accumulate authentic, objective factual knowledge in science, is that scientific results and theories are assessed solely with respect to their adequacy to the facts, the justice that they do to observational and experimental data. Human feelings, desires, problems, aspirations, values, must be ruthlessly excluded from all consideration when it comes to the assessment of scientific results. Only in this way can science hope to achieve genuine objective factual knowledge. Thus to this extent the very integrity of science depends upon a sharp split being maintained between the scientific, intellectual, rationalistic domain, and the domain of human feelings, desires, values. It is, of course, the case that scientists pursue scientific research as a result of all kinds of "extra-scientific" human desires and motivations. Scientists may pursue research out of a passionate sense of wonder, passionate intellectual curiosity; they may desire to discover that which will be of genuine value to suffering humanity; they may wish to win recognition and admiration from colleagues; they may wish to create something that endures; or they may desire to advance an academic career. The all-important point however is that when it comes to the assessment of results, all such extra-scientific, extra-rationalistic human aims and motivations must be put entirely on one side, and justice to the facts of experience alone be taken into account. When it comes to the assessment of results one aim alone can be entertained: to discover Truth. The success of science, the objectivity and authenticity of scientific knowledge, depend crucially upon this severe discipline being maintained. And indeed the ultimate value of science to humanity depends upon this kind of dissociation between the intellectual aspects of science, and the human aspects of science, being preserved. For science is only of value to humanity to the extent that science is able to produce authentic, objective factual knowledge. Thus, in order to be of value to people, it is essential that science *ignores* the interests,

11

feelings, desires and values of people, in the kind of way just indicated. Assessment of scientific results – the heart of scientific method – must inevitably be a specialized, technical, intellectual task having nothing to do with the feelings, desires and values of people: no wonder, then, that people who are not trained scientists can only rarely have a contribution to make."

Admittedly, not all individual professional scientists would probably reply in exactly these terms. Nevertheless, the argument just outlined does, I believe, represent a broadly agreed view of science which is in practice upheld by most scientists, and which is to a considerable extent *embodied*, as it were, in the institutional set-up of science as it exists at present. Without serious injustice, we can, I think, treat the above as reasonably accurately representing the views, the standards, the intellectual values, of the Scientific Mind.

Now the crucial point to notice about the above argument is that it defends and justifies the dissociation of science from human life in terms of a certain *rationalistic ideal* for science, a certain *philosophy* of science, a view concerning the proper aims and methods of science. The cardinal points of this viewpoint might be summed up like this. The fundamental intellectual aim of science is to improve our knowledge of value-neutral matters of fact. This aim is achieved by assessing results solely in the light of observational and experimental data. This, in turn, requires that the scientific, intellectual domain of science be dissociated from the domain of human needs, desires, feelings, problems, values and aspirations – the domain of human life.

Let us call this rationalistic ideal for science, this generally agreed philosophy of science, this widely upheld conception of the proper aims and methods of science, *standard empiricism*.

Now, quite clearly, if standard empiricism does represent a truly rigorous, rational ideal for science, then it is going to be very difficult to develop the kind of person-oriented science indicated above. Requirements of scientific rigour, objectivity and intellectual integrity are, as it were, at odds with our human desire to develop a science sensitively and intelligently arranged around human life, sensitively and intelligently designed to help us enhance the quality of our lives. As we try to develop a rather more person-oriented science, so, alas, we are almost bound progressively to undermine the rigour, the objectivity, the intellectual integrity of science, the authenticity of scientific knowledge. It was just this, indeed, that the above argument of the Scientific Mind was designed to establish.

The fundamental idea of this book can now be put like this. Standard empiricism, far from representing a rigorous, rational, objective ideal for science, very seriously *fails* to capture, and do justice to, the real rigour, rationality and objectivity that is inherent in science at its best. Standard empiricism represents a *false* ideal for science, a *false* philosophy of science, an unacceptable idea of what it is to be scientific. The above argument of the typical scientist designed to defend the dissociation of science from life is very seriously invalid, resting as it does on a seriously inadequate philosophy of science. *As we develop a more truly rigorous conception of scientific enquiry, and as we develop a more explicitly rigorous science, then we will find precisely that we are developing the kind of conception of science, the kind of science, that I have called "person-oriented".* Science today fails very largely to correspond to what I have called person-oriented science just because the attempt has been made, on behalf of the scientific community, to keep science confined within the straitjacket of the seriously unrigorous, irrational conception of science of standard empiricism,

In short, rationality and human desirability walk hand-in-hand together. Person-oriented science which may, to the professional scientist, at first sight, seem to confound scientific rigour, rationality and objectivity, actually represents the finest flowering of scientific rigour, rationality and objectivity. The demand for a science that is more sensitively and intelligently responsive to the needs, desires, problems and aspirations of human life is at one and the same time the demand for a science that is more rigorous, rational and objective. As we intensify intellectual standards, as we make science more exactingly scientific, more fiercely rigorous and objective, so we discover – lo and behold – that we have given birth to *people's science*, to something that can proudly stand as our version of the Pygmies' songs.

And we can go further. It is not just science that becomes more humanly desirable as it becomes more intensely, rigorously, *scientific*. The same thing goes for *reason*. As we make our ideal for reason more truly rational, more rigorous, more intellectually exacting, as it were, so we find that what we develop becomes increasingly *desirable* (and especially so for those who are inclined to think of reason as somewhat undesirable). Indeed, we discover that reason, properly understood, is simply that which helps us to discover and attain the desirable, that which is of value. Reason is desirable because reason helps us to realise desire: and if reason clashes with desire, then it is our ideal of reason that is at fault (or our use of this ideal). It thus runs counter to the whole spirit of reason to try to use reason to force people to accept unwelcome truths. Far better – far more rational and desirable – to use reason to encourage people to discover for themselves what is for themselves truly desirable. A rational argument is an aid for the disclosure of hidden beauty, obscured delight.

Thus reason, properly used and understood, does not in any way place *prohibitions* on thought and action. It does not in any way *impose* itself upon us. On the contrary, its whole purpose is to enhance our capacity to choose and act as *we* really desire: it is designed to help us enhance our freedom.

In terms of this new, desire oriented ideal for reason, it suddenly becomes possible to hold that the truly rational life, the truly rational society, is profoundly *desirable* (almost by definition, as it were). Reason is our finest song to Nature to help her to help us to attain our hearts' desires. Reason is a beautiful key in which to sing our songs to each other.

During the course of the next eight chapters, I hope to lay bare something of what seems to me to be the hidden beauty, the partly obscured desirability of these suggestions.

CHAPTER THREE

AN ANGRY CLASH BETWEEN SCIENCE AND PHILOSOPHY

A problem of procedure and presentation now arises. My concern is to *show* that scientificness and human desirability walk hand-in-hand together, mutually enhancing each other; and as an aspect of this, that reason and desire go hand-in-hand together, again, mutually benefiting from each other.

The idea is not a familiar one. It has perhaps a certain novelty, a paradoxical air. By and large, it is not an idea that I have found very easy to communicate to colleagues and friends.

A part of the trouble is this. I believe the idea is true, important, beautiful, useful. I therefore feel that I must assemble all the vast array of powerful arguments that I *know* go to support the idea. I must make these arguments utterly compelling, so that academics and scientists especially will be *forced* to take the idea seriously. All objections will be silenced. The absolute decisiveness of my arguments will compel people to concede the point: Yes, science and reason, properly understood, simply help us to discover, to experience, to know, the desirable, the beautiful.

You can see the trouble. A certain tension begins to emerge between the message and the manner of its delivery (to put it mildly!). Whatever else one may mean by "rationality" it can hardly be "rational" to be as self-contradictory as this. And even if the idea is a good and beautiful one, this is not likely to be noticed: if one senses one is being *forced* to notice the beauty of an object, one is hardly likely to see it, feel it. The "inconsistent", powerful argument is almost bound to fail.

In fact, in writing this book I sailed right into this trap, with my usual impetuous stupidity. The more cogent and potent my arguments became the more *useless* they became. I rewrote, rewrote, rewrote, striving to strike the right balance: but it was no good. The essential thing was being missed.

15

On the other hand, how am I going to get through to the scientific mind, to be *heard* at least? It's no good saying "wouldn't it be lovely to do physics the way the Pygmies sing their songs". "A lovely idea", will come the response. And everything will go on as before. And I really do believe that there is a *need* for a big change, that it really is desirable to make a big change. But in order even to make the suggestion in a way in which it will be noticed, somehow one has to enter the ferocious intellectual field of force of academic science, held rigid by fierce intellectual tension. Intellectual *dynamite* is needed, simply in order to create the effect of clearing one's throat before beginning to speak! A dazzling display of *fireworks* is needed, in order to induce scientists and academics to look up momentarily from their lab benches, experiments, equations, journals, lectures, conferences, habits of thought, simply in order to see, just for a moment, what is going on. Pygmies' songs, or their equivalent, do not, I am afraid, have sufficient intellectual potency, as these things are at present judged, to receive even momentary attention or notice, let alone serious thought!

It all comes from seeking to change, to improve intellectual standards, ideals for science and reason. If you set out the case for the need to develop new improved ideals in terms of the old ideals, inconsistency must arise between style and content. If you set out the case for a need for change in terms of the *new* ideals, you will be consistent: but you will not be heard.

A dilemma indeed! It occurred to me that it might be possible to overcome this expositional problem, to some extent, by means of the following dodge. Let us assemble here, on these pages, a little band of puppets, to argue it all out among themselves, for us! Let there be a puppet standing in for the conventional intellectual standards, conventional ideals for science and reason. And let there be a puppet who believes passionately in Pygmy science. Let the dilemma be *his* dilemma: we can watch how he makes out. These puppets can be permitted their own voices, their own styles of reasoning, arguing, quarrelling, their own styles of gesturing in the direction of what seems to *them* to be desirable. And we can watch: *my* responsibility will be to try to see that justice is done.

And of course it is entirely up to you, yourself, to decide what you wish to make of all this – so the overall exposition ought, ideally, to match the overall message.

And at the same time we can have our dynamite, our fireworks! Only it will be *emotional* dynamite, *dramatic* fireworks. The driving force can be feeling, emotional conflict, desire and fear, rather than *logic*. The logic can itself be gossamer thin, as delicately responsive to will and desire as anyone would wish for.

Without more ado, let us begin. Let me introduce our first two puppets, who we may call *Scientist* and *Philosopher*. (If they seem a bit wooden, well, please remember, they are only puppets – Petrushka, Pinocchio intellectuals, themselves sad, no doubt about their rather wooden minds, hearts and lives. They would like to be people!)

Both are in their mid-thirties. Both are serious, intelligent, sensitive, dedicated men. The scientist is married with two children; he leads a fairly settled, steady way of life. He enjoys classical music, French literature, football, good beer, and is, almost despite himself, deeply concerned about our troubled times – though he feels, like the rest of us, his own personal powerlessness actually to *do* anything to help. His scientific work is fairly technical; I forget exactly what he does. It could be theoretical physics, cosmology, molecular biology, neurology, embryology – even psychology, perhaps, of the somewhat more severely experimental, scientific kind. In any case, his research is important, exciting; he has already achieved one or two results of some value, acknowledged and accepted by those working in his immediate area of research. His colleagues respect him, professionally. Despite frustrations, set backs, a certain impatience with what he feels to be harmful, foolish, intellectual and financial restrictions in the scientific world, he nevertheless feels himself to be extremely lucky in having captured a worthwhile job that also interests and fascinates him. He is all too well aware that this is in marked contrast to the fate of most men and women alive today.

One aspect of our times, especially, disturbs him. It is what he would call the cult of unreason, a certain fashionable disparagement of science, fact, logic. For him, science, and the scientific attitude, are of supreme value. They constitute a kind of lifeline to sanity and decency in an often insane world. Science, for him, constitutes our very best attempt to know and to understand the world, with complete objectivity and impartiality, unaffected by our feelings, our desires, our hopes, our dreams. Throw away *scientific* knowledge and understanding and we throw away our knowledge and understanding of objective reality. We will begin to believe fantasies, hallucinations. A society without scientific knowledge is like a person without commonsense knowledge – a person living in a dream, a mad person. Without science, we have lost our lifeline to sanity and to reality: horrors will be unleashed. Modern science, in short, is for him something like a foundation stone for civilisation. Probably he would not express himself in quite these terms and certainly he does not carry such thoughts around with him consciously from day to day. But something of this kind represents his deepest underlying convictions and feelings about science.

Our *philosopher* in contrast to the above, leads a much more unsettled, emotionally chaotic life. For him, there cannot be the neat separation of intellectual life and personal life that holds for the *scientist*. His concern, absurdly, is to understand reality, physical reality and human reality (and the inter-relation between the two). At times it is almost as if he regards his own life as a kind of test case, an experiment, against which he can test his philosophical ideas. "Experience" for him has a quite different meaning from that of the scientist. It means *personal* experience – the whole human range of feeling, perception, desire – not detached, impersonal observation and experimentation of the scientist.

His fundamental intellectual preoccupation is with relationships between people – and especially with the nature, the problems, of the most desirable, valuable form of love, whatever that may be. At times it has almost been as if he has plunged into relationships solely in order to further his intellectual, philosophical enquiry into the nature of love. His relationships have almost been the raw material in terms of which he can make his discoveries. To his

18

horror and confusion, he found that this thirst for understanding had led him on occasions to act in what seemed altogether a heartless fashion. Or was he using his concern to understand as an excuse for heartlessness? He did not know. His concern to understand lost some of its conscious character, abandoned out of guilt. But it nevertheless lived on, despite himself, a constant worrying questioning, probing. He became aware of the absurdity of putting *intellectual enquiry* before *life*. How are we to live? How am I to live? What is there that is best in life to pursue, to attain? And how does one set about realising what is best? These are his fundamental questions, his fundamental concerns, and at times the very fact that he has taken these questions so seriously has seemed like the source of the trouble. Does it not betray a deep anxiety, a lack of trust and confidence? Thought used in order to discover how to *begin*, rather than being used in order to discover how to *improve* that which already exists: what could be more absurd, more irrational?

Gradually it began to dawn on our philosopher: it was not all his fault. His troubles were in part *objective* – due to defects in the very ideal of rationality and intellectual enquiry which he was trying to use in order to discover how to live. Just because he had absurdly combined the two questions: How am *I* to live? and How are we all to live? (or what is of value for *me*? and what is of value for all of us?) he personally became terribly vulnerable to defects in public concepts of reason. A slight defect in the notion of reason he was attempting to employ would plunge him into despair for months, even years. And he would interpret the trouble as entirely *his* fault, failing to see that there was another aspect to his problem. His own deeply personal problem was not only his problem: others too, in different ways, were struggling with similar difficulties. The heartlessness of his interest in the nature of love, for example, really did in part arise out of a kind of heartlessness inherent in intellectual enquiry in itself. Precisely this heartlessness of intellectual enquiry, this lack of sensitivity to personal feelings and desires, actually prevented people from relating thought to experience, and experience to thought. And so gradually our philosopher came himself to develop a viewpoint which is actually quite close to the viewpoint that I wish to advocate in this book.

(Our puppet philosopher incidentally thinks he has it all worked out: *that* is always his biggest mistake.)

Having made what seemed to him to be a beautiful discovery, which ought at least to be of *interest* to others, he found the greatest difficulty in conveying to others what it was he thought he had discovered. In particular, there seemed to him to be no hope of interesting his academic colleagues, who in any case had a tendency to regard him as a somewhat irrational, unbalanced, unscholarly, intemperate academic. His situation could not be more different from that of our *scientist*. In complete contrast to our *scientist*, our *philosopher* was, in essence, on his own, unable to collaborate with others, unable to communicate anything but odd scraps of his discoveries (twisted by academic convention and assumption to something remote from his original intentions). He knew very well that his colleagues simply derided him: there was nothing like *respect*. One contemporary philosopher only, it seemed to him, pursued philosophy in something like the same spirit: namely, Karl Popper. And yet all communication with Popper proved impossible.

A mismatch had come into existence between the objective world and the philosopher's private, personal world. Either the world was mad, or *he* was mad. Gradually, he worked out a more balanced, and productive perspective. It no longer seemed that the one fundamental thing that he had to communicate to others, to his friends, and anyone else interested, was what seemed to him his great discovery – the secret of life, just that which he had always wished to discover. Life is of greater value than any secret, any message or methodological discovery. Did not his methodological discovery assert precisely this point?

Here then are our puppet protagonists. Both are vulnerable men. Both have sincerity, honesty, intelligence and passion – although whether anything other than a display of passion will be on show *here* remains to be seen.

Both have what is perhaps very characteristic in men today, a kind of troubled male pride (vulnerability experienced as a wound in masculinity rather than its finest expression, male *courage* being needed to acknowledge and express vulnerability). In the case of our scientist and philosopher, all this is transferred onto the intellectual plane. Both have invested much of themselves in their intellectual work. A certain outward intellectual arrogance acts as a buffer for an inner vulnerability. This makes rational discourse difficult when basic issues and assumptions are at stake. In fact, if it were not for my own personal intervention, our protagonists would quickly cease to be on speaking terms with each other.

They meet. They talk.

SCIENTIST: I would like to begin, if I may, with a statement. It is all very well to come along with some piece of quasi-fashionable philosophy which asserts that science as it exists at present is unrigorous, irrational. I would like to remind you of the immense success and stature of science. It is beyond all doubt: science during the last two hundred years – or even during the last fifty or twenty years – has profoundly deepened and extended our knowledge and understanding of the world around us. Some of the greatest minds produced by the human race have devoted themselves to the furtherance of science. In countless ways science, through technology, has utterly transformed the human condition for the better. There is simply no comparison whatever between life in Medieval England and life in England today. Dangers, abuses, of scientific knowledge there may be: but the fault can scarcely be said to lie with *science*. In itself, knowledge is morally neutral; it is what we *do* with it that counts. And by and large scientists themselves have been as actively concerned with avoiding the dangers of ill considered or immoral technology as much as anyone else. I . . . . .

PHILOSOPHER: I am in agreement with almost everything that you have just said. In fact I think I would want to go even further in emphasising just how magnificent is science at its best. Popper once called science one of the great spiritual adventures of mankind, and . . . . .

SCIENTIST: Could I possibly finish my opening statement?

PHILOSOPHER: Of course. I . . . . .

SCIENTIST: The point that I intended to stress is simply this. There is, in my view, an element of absolute absurdity in suggesting that there is something seriously wrong with intellectual standards as exemplified in scientific research as it exists today. Incompetent scientists there may be. Shoddy scientific work there undoubtedly is. But to suggest that there is something seriously wrong with the whole intellectual framework of science as it exists at present indicates, in my view, something approaching a touch of insanity. I am here because I have been asked to come along. For myself, I do not imagine that our discussion will be very profitable.

PHILOSOPHER: Thank you. I feel really encouraged to go on.

SCIENTIST: You have no choice.

PHILOSOPHER: I know.

SCIENTIST: Perhaps you could begin by outlining, in extremely simple, non-technical terms, what it is you want to say.

PHILOSOPHER: (Heaves a sigh.) Very well. I'll do my best. It can be summed up like this. There is, to begin with, human life, people pursuing their goals, seeking to give and to receive love, share happiness, overcome problems, cope with appalling conditions of near starvation, political and economic enslavement. . . . .

SCIENTIST: Yes, yes, yes. You may be surprised to hear it, but I am actually quite as well aware of all this as you are.

PHILOSOPHER: I am seeking only to set a context for what I have to say. Next, there is *science*, the personal, social, institutional, intellectual reality, science as it actually goes on in laboratories, lecture halls, seminars, scientific journals, scientific monographs, technical scientific books, committee meetings, popular books on science, examinations

SCIENTIST: Again this is all *fairly* familiar to me.

PHILOSOPHER: (Our noble philosopher decides to ignore this sarcasm.) And finally there is the *philosophy of science*. My *thesis* amounts to this. The scientific community, in a somewhat unthinking, carefree fashion, has come to accept a certain *philosophy of science*, which I call standard empiricism. This philosophy represents, as it were, the official conception of science, the official ideal for scientific integrity, rigour, objectivity, rationality.

In essence, standard empiricism is extremely simple. It asserts: the main aim of science is to improve our knowledge of value-neutral factual truth. In order to do this, theories need to be assessed impartially in the light of experimental success. Scientific integrity demands that personal, social desires, feelings, needs, values, hopes, fears, dreams, be ruthlessly excluded from the intellectual domain of science.

In largely unnoticed ways, this carelessly accepted *philosophy* of standard empiricism influences scientists in what they *do*. It influences such things as what can be published in scientific journals, under what circumstances ideas and results are accepted and rejected, what problems are tackled in research, scientific education in schools and universities, scientific textbooks, both as regards style and content. But of far greater importance than all this, the philosophy of standard empiricism exercises a profound influence over the whole way in which *science* is related to *people*, to *life*, to *society*. Standard empiricism contains within itself an idea of how science and people ought ideally to be related; and scientists, in good faith, do their best to ensure that this relationship is, in fact, established and maintained.

The fact is, however, that this widely accepted and profoundly influential philosophy of standard empiricism – stretching its tentacles far and wide into both science and society – is an absurdity, a piece of nonsense, complete rubbish. And worse: it is *harmful* rubbish, restricting, destructive rubbish. Standard empiricism is a kind of straightjacket, into which we thrust science (and ourselves) convinced that it is all for the best, that this is the only way to have a truly rigorous, objective, *scientific* science. And it is all the other way round. If only science were to be liberated from this appalling straightjacket, it could flower, flower into something of unguessed beauty. We would have Pygmy science on our hands. Science would be our beautiful songs to our beautiful world, to help our world become more beautiful. This would be everyday common sense, for everyone; and science would be experienced in this way. At present, science is not given the opportunity to be the beautiful thing that it could be. And, as a result, it is not just *science* which suffers; *we* suffer; *we* suffer.

SCIENTIST: You do, anyway.

PHILOSOPHER: (Excitedly carried away now with what he is saying) Please. Do please let me have my say: then you can give me your comments.

To continue with my exposition.

The widely accepted and profoundly influential philosophy of science of standard empiricism is defective in at least the following ways. It completely *misrepresents* the main aims of science. Science does not seek factual truth as such. Science seeks *valuable* truth, *important* truth, *beautiful* and *useful* truth. And further: science seeks valuable truth in order to be of help, of value to people. And further still; the heart of scientific *method*, when properly understood, is something which can be generalised to form a new desire oriented concept of reason, of universal relevance and value for *all* our personal social pursuits, problems and activities. And finally, science is not something pursued exclusively by experts, by disembodied minds: fundamentally and centrally it is *our* enterprise, an outcome of our concern with our lives, the world around us, each other. Science helps us to develop better relationships with the world, and with each other. Life, human action, comes first; and science, culture, rationally understood, is delicately, sensitively arranged around our actions, our lives, to help us develop our actions, our lives, in fruitful directions, in directions which we really want.

SCIENTIST: (Ironically) I see.

PHILOSOPHER: (Not noticing the irony.) You do? You really do? But wait. Just a moment and it will be finished.

The *consequences* of the fact that standard empiricism completely misrepresents the real aims of science are disastrous. *First*, it becomes impossible to make rational sense of even the most elementary things about science, such as, for example, how it is that things can be "verified" or rationally chosen in the light of evidence, of experimental success. *Second*, and far worse, standard empiricism upholds methods, official rules of procedure, that are completely stultifying. Those scientists who have dutifully sought to realise the official aim of improving knowledge of value neutral factual truth have achieved little: all the great achievements in science have been made by men and women who cared too passionately to discover that which is of *value*, to bother about observing the niceties of official scientific propriety (apart from making certain concessions when it came to *publication*). It is precisely this which accounts for the fact that *discovery* in science is widely held to be an irrational, or extra-rational, process. Relative to official concepts of "reason", scientific rationality, discovery is indeed "irrational". The fault lies, however, with conventional standard empiricist conceptions of rationality, not with what goes on in creative discovery in science.

Thus, as a result of quite fundamentally *misrepresenting* the basic aims of science, standard empiricist methods and procedures for science are actually *obstructive* rather than helpful. As a result of misinterpreting the proper aim for science, the proper direction for scientific enquiry to take, standard empiricism, not surprisingly, is worse than useless, a hindrance rather than a help. Progress in science has been achieved *despite* official, institutional, acceptance of standard empiricism, not *because* of it.

But we have not yet come to the worst of it.

SCIENTIST: I am sure we haven't.

PHILOSOPHER: Sh! The *third*, and infinitely the most damaging consequence that flows from the widespread attempt to make science conform to standard empiricism is that a vast gulf is set up between the objective intellectual domain of science, and human *life*, feelings, desires, frustrations, hopes, fears. Science becomes insensitive to people: and people become insensitive to science. Our minds and our hearts become divorced from one another.

Scientific, intellectual problems cannot be understood and tackled intelligently, as aspects of human, social problems, our concern to solve a scientific, intellectual problem being a part of our concern to help solve human, social problems. The possibility

25

of delicately, sensitively arranging our science, our reason, around our *lives*, so as to encourage the flowering of our lives, disappears. The proper human use and value of science is lost sight of. And all this happens because the ideal of scientific rigour, accepted so carelessly and thoughtlessly by the scientific community, actually insists that if science is to retain its integrity, its objectivity, a sharp rift *must* be maintained between the scientific and the social, the intellectual and the personal, mind and heart, reason and desire, thought and feeling.

In brief, this misconceived philosophy of science of standard empiricism (i) fails completely to make rational sense of science, (ii) serves, if anything, to *obstruct* rather than *promote* scientific progress, (iii) utterly disrupts, dislocates the delicate, harmonious, and humanly valuable relationships that ought ideally to exist between science and people.

The absurdly misguided attempt to make science conform to the (false) ideal for scientific rigour of *standard empiricism* is, in short, in essence what prevents us from having a truly *rational* person-centred science.

SCIENTIST: So it is really all the fault of the philosophers?

PHILOSOPHER: Yes! Exactly! Most *scientists* know in their heart of hearts that standard empiricism is an absurdity. What they ought to conclude from this is: "If our *philosophy of science* is seriously defective, if we fail in some way quite fundamentally to understand properly this highly sophisticated, intellectual enterprise of science – then perhaps science itself suffers. Perhaps our *practice* of science is defective. Or at least, there might well be substantial room for improvement, for an enhancement of rigour having, for a change, fruitful *scientific* consequences (much as the work of Weierstrass and Dedekind in the nineteenth century, in making the calculus more rigorous, and more comprehensible, made important contributions to mathematics itself.)" But alas scientists do not draw this *obvious* conclusion from the blatant inadequacy of customary philosophies of science at all. On the contrary, they conclude that the whole topic is merely a waste of time, sterile. Science is something you *do*, and know instinctively how to do, like a craft or a practical skill. All your philosophising, your thinking about the nature, aims and methods of the activity is a waste of time, an occupation for fools and charlatans.

And this typical attitude of scientists is only further strengthened by what goes on in the philosophy of science. For the last two hundred years or so – ever since the staggering success of Newtonian science became generally acknowledged – philosophers of science, epistemologists, have struggled with one basic problem: How can one understand the *success* of science? How is knowledge *possible*? (As Kant put it.) Philosophers have not sought to help science to be more successful. On the contrary, it is the achieved success of science that creates the problem. In all this time they have failed to understand the absolutely *elementary* point of how science could possibly have achieved its success (if only science was not so successful philosophers could, as it were, relax). Nothing could indicate more dramatically just how unbelievably inadequate philosophers' understanding of science has been. After all, we usually set out to improve our understanding of some activity in order to help us do it even better. If, after two hundred years of sustained effort, we are still no nearer understanding the success that we, ourselves, have achieved then we *must* be thinking about our activity in a quite disastrously defective way. Some simple, elementary blunder must be being made.

Did philosophers draw this obvious conclusion from the absolute failure and sterility of all their attempts to make rational sense of science? By and large, no! They did not kick standard empiricism out of the window, as a result feeling free to develop a different, more adequate, accurate and fruitful philosophy of science. Quite the contrary, they redoubled their efforts to prop up standard empiricism. They encrusted standard empiricism with a thick morass of academic complexity and irrelevancy. The problems grew more and more technical, more and more difficult, more and more remote from anything that could possibly interest a scientist. And as a result of this Herculean and utterly useless labour, it became more and more impossible effectively to make the simple point: look, if all our attempts to make rational sense of science

27

within the framework of standard empiricism fail, then perhaps it is the framework that is at fault. Perhaps science doesn't seek value neutral truth *as such* (as almost all philosophers, in one way or another, underneath their subtleties presupposed). Perhaps science seeks *valuable* truth, humanly *desirable* truth. Perhaps indeed science is simply a part of our common human endeavour to discover and realise beauty, that which is of value, the good in life. Perhaps at its best science is the rational outcome of human desire, a part of our rational attempt to realise the humanly desirable. Perhaps our inability to understand science means there is something seriously wrong with the whole way in which we are *thinking* about science, and even *doing* science.

As a result of the failure to ask these obvious questions, and consider these simple and obvious possibilities, philosophers produced almost nothing of any real interest or value concerning science (it is to the writings of *scientists* one turns to get insight into science). And this, of course, only confirmed scientists in their suspicion that philosophy is a useless, vacuous activity. They *should* have concluded from this long-standing grotesque failure of philosophers to make rational sense of science: "Oh, something serious is wrong. The issue is much too important to be left to these incompetent fools. We had better ourselves take a hand in sorting the thing out." In fact they tended to conclude: "So, philosophy of science is a useless activity of no practical value to scientific research itself". Beveridge, in his delightful and instructive book, *The Art of Scientific Investigation*[1] (a kind of instruction manual for the young research worker), justifiably feels no need whatever to refer to philosophy of science, or the study of scientific method. (Perhaps the one great exception to all this is Einstein, who once wrote, with his usual total perceptiveness: "Epistemology without contact with science becomes an empty scheme. Science without epistemology is – insofar as it is thinkable at all – primitive and muddled."[2] But then Einstein alone among scientists and philosophers *rejected* standard empiricism – in many ways the key to his great scientific success.)

Basically all that philosophers of science have achieved during all this time is to help maintain and preserve the central rationalistic neurosis of science.

SCIENTIST: So, science is neurotic now, is it?

PHILOSOPHER: I mean it quite seriously. As we shall see later on we can reinterpret the psychoanalytic notion of neurosis so that it becomes a rationalistic or methodological idea, a methodological complaint which any aim pursuing enterprise or entity may fall into. If some enterprise seeks to realise some desirable but highly problematic goal, A, rejects the idea that it is seeking A precisely because of the *problematic* character of this aim, and instead declares to itself that it is seeking the ostensibly unproblematic aim B, while all the time continuing surreptitiously to pursue A under a cloak of rationalisation, then that enterprise may be said to suffer from "rationalistic neurosis". In these circumstances, clearly, official rationality or philosophy becomes worse than useless. The more honestly and "rationally" B is pursued the more unsuccessful will the enterprise be (from the standpoint of realising A). Only a highly dishonest, hypocritical, irrational pursuit of B will achieve *real* success (namely, realisation of A). For those who think B really is the proper aim of the enterprise, precisely the *success* of the enterprise will seem most puzzling. Science is just one example of all this.

SCIENTIST: (Very dryly) I suppose science represses its unconscious desires?

PHILOSOPHER: (Still noticing nothing, carried away with his enthusiasm) Oh! You see it! I am so delighted. Exactly. The real, but profoundly problematic goal of seeking to improve our knowledge of humanly *valuable* truth is suppressed, just because values seem to be too problematic to cope with within a scientific context. This aim is replaced by the apparently innocent, unproblematic (but actually nonsensical) aim of seeking to improve our knowledge of factual truth *as such*. Scientists *in fact*, entirely sensibly, contrive to seek to discover valuable truth. Philosophers, convinced that science ought ideally to seek truth *as such*, fail completely to understand the success of science. Philosophers spell out methodologies appropriate to realising truth *as such*; not surprisingly, these methodologies are singularly inappropriate and unhelpful for realising the *real* aims of science. Philosophers try to make sense of what scientists *say* they are doing: and they are disheartened when scientists ignore all the results of their labours. The whole thing would be a comedy of error, were it not for the truly harmful and tragic *consequences* of all this confusion, when judged in human terms.

SCIENTIST: Science seems to have done quite well for itself despite its neurosis. Perhaps in the circumstances "neurosis" is not such a bad thing to have?

PHILOSOPHER: Fortunately, in their actual research, to a considerable extent, scientists ignore the "official" conception of science of standard empiricism. But science still suffers nonetheless, from its misrepresentation of aims, both in intellectual terms and in human social terms (the two failures being in essence two sides of the same coin).

SCIENTIST: Well, if you want my personal reaction to all this, it is this. Neurosis may well be around in the air somewhere, but I doubt that it is to be associated with *science*.

PHILOSOPHER: What do you mean? (It begins to dawn on him just what it is that the scientist does mean.) Oh, I see. You think all this is just the raving of a lunatic.

SCIENTIST: *You* said it.

(There is a long awkward pause.)

PHILOSOPHER: (In bitter, angry, hurt, unphilosophical tones) And would you like to know what I really think?

SCIENTIST: Fire away.

PHILOSOPHER: You tell us you care about rigour, objectivity, intellectual standards, scientific integrity. These are things you care about, isn't that right?

SCIENTIST: Yes.

PHILOSOPHER: And you are inclined to deplore and condemn what you see in our times, people who allow feelings, passions, personal impulses to influence and guide their thoughts?

SCIENTIST: On occasion, yes.

PHILOSOPHER: Well, then, my God: I say this to you. Put your own house in order! Put your own house in order before you go out condemning others. Your precious intellectual integrity is a shambles. You have almost stopped thinking – except in highly stereotyped, technical, fixed ways. You who claim to prize reason, practice unreason, and you are too proud, too arrogant, to acknowledge your elementary blindness. If it was just some academic discipline at issue, the thing would not matter. But *people* are caught up in all this. Human lives are entangled in your institutional, intellectual neurosis. Science stands at the centre of our modern world, too important and influential to be ignored. You have a heavy responsibility to others. It is time you swallowed your own petty intellectual self-satisfaction, and took a closer, more honest, more self critical look at what is really going on. It is quite possible that we shall succeed in destroying ourselves some time during the next decades. I can assure you, in my view, that if we do, precisely the intellectual arrogance of scientists will have made its contribution. We live in neurotic times; and that fact is not unconnected with the central rationalistic, methodological neurosis of science.

SCIENTIST: (Abruptly) Well, let's leave it an open question as to who needs a psychiatrist. I have had enough. I am off.

(They separate.)

CHAPTER FOUR

WHAT'S WRONG WITH PHILOSOPHY?

In ordinary circumstances, this one experience would probably be sufficient; our two protagonists would not again talk together about these issues. I, however, as a *deus ex machina*, contrive to keep things going.

In literature we demand that emotional conflict be resolved one way or another, so that we may experience the catharsis of emotional resolution. In life, on the other hand, it rarely seems to happen like that. Our angers, frustrations, irritations, resentments, fears, bitter jangling feelings, are left hanging in the air, unappeased, unresolved. There is, of course, a popular theory that one has a simple choice: either *express* feeling or *repress* it (which means retain it, bottled up within, to do further damage). Our philosopher believes that there may be a third option; to abandon unhelpful feelings, laugh them away, blow them into the wind; recapture a poise, a balance, a serenity. At any rate, this third possibility seems to him worth working for: for if it can be achieved it means one can be free, open, careless, spontaneous with one's feelings, letting them rush up, soar, plunge, dance as they please. But when they go wrong, when they become unhelpful, destructive, useless, then one does not have to *repress* them, thus building up a reservoir of unacknowledged anger (which, of course, ruins everything): one can blow them away, laugh them off. Even passionate, overwhelming dramatic feelings can be taken lightly, gracefully, in no way a threat to one's intelligence, one's perceptiveness, one's objectivity, one's freedom of action. This, at any rate, is his *theory*: he does not always succeed with the *practice*. (He finds, in fact, that he is singularly inept at the practice.) I, however, the author, can arrange these things with the greatest of ease (as far as our protagonists are concerned). Concerned for the welfare of my argument (and of course not indifferent to the welfare of my puppets) I blow a gentle puff of liberating good humour through our poor philosopher's troubled wooden heart and mind.

His situation is, for him, so paradoxical, so contradictory. He feels at times that he is caught in a kind of trap. He has, he believes, perceived an idea, a possibility of great beauty. In a rational world it would be sufficient to rush up to people and say: Look what I have seen! Delicately, sensitively, perceptively, the treasured perception would be explored, played with, laughed over, developed. After sympathetic consideration it might turn out to burst like an iridescent soap bubble, yet another delightful absurdity. And that would be all part of the game. How else are we to discover objectively beautiful possibilities, if we do not explore, develop, expand, with the film of our imagination, our paradoxical fantasies, our momentary dreams and visions? Bits and pieces of past broken dreams may be fitted together to form a new realistic dream – which might never get to be built without the aid of the fragments from the old broken dreams. The very activity of optimistic, light hearted, sympathetically critical dream bubble blowing is desirable to encourage. An openness to this activity of bubble blowing – especially when performed by people who have values, ideals, aspirations, different from one's own – can enable one to discover ways in which one has misrepresented to oneself one's basic aims. If *science* promoted, instead of positively discouraging, bubble blowing, the misrepresentation of aims at present built into the institutional structure of science – as seen by our philosopher – would soon vanish away. The institutional neurosis would be cured. Surely a person should never be put down for indulging in such an activity, and in seeking to pursue this activity with others in a light hearted fashion (unless, of course, there are more immediately important things that need to be done on hand. Life, actual life, comes before bubble blowing – which is only at most a part, a bit, of life.)

And what was this beautiful idea, this beautiful possibility, this iridescent bubble that our philosopher had discovered, or blown, which he wished to share with others? Very simple! The delight-fulness, and the value, of bubble blowing! This, in essence, is what *reason* is all about – unravelling, exploring, scrutinising our desires, our dreams. This is what should be going on in science openly and delightedly – if science is to be pursued in a truly rational fashion. This is what *education* should be encouraging children to do. And this is what we should be doing in our lives, as a part of our various personal, social, institutional, political activities and enterprises. It would be delightful to do this; and it would be *useful* to do this; for we could, as a result, enormously enhance our capacity to discover and attain that dream which is the most desirable, valuable and realisable, that dream which does the greatest justice to what is best in all our individual, personal dreams. Bubble blowing, if pursued in a carefree and critical fashion, could help us solve our problems, attain our most desired, valued human ends.

This is our philosopher's bubble! And now comes the trap, the paradox. Our philosopher finds he can only communicate the idea when the idea has *already* been understood. It is only communicable when it does not need to be communicated.

The rest of the time he finds that his attempts to communicate his bubble creates only tension, trauma, anger. It seems to our philosopher that people put too much of their own identity into their dreams, their ideals. They take their dreams, their ideals, too seriously, almost more seriously than life itself. And as a result they prevent themselves from *learning* as they live; they obstruct the possibility of exploring possible dreams, easily, carelessly, critically, so that such an activity may both be a *pleasure*, a delightful game, and something that might have a practical value, a use for life itself.

It seemed to our philosopher that all too often people "found" themselves in their dreams, their ideals, their aspirations: they found also that others holding different positions, did not appreciate their own dream, their own precious identity. Thus, these others became a threat, endangering one's own identity, one's own soul. Like minded people gathered together in tribal groups to give each other support and confidence, further "fixing" the common dream. The others are held to be madmen, fools, idiots, poor benighted lost souls, In this way, people become bewitched by their dream. Instead of people choosing dreams, as a result of

34

playful bubble blowing and bubble bursting, dreams choose people, and lead them remorselessly by the nose. Zestful, imaginative, ever self-renewing children grow into haunted, trapped, weary adults. The lost freedom, the abandoned play, only increase the frustration, the anger, the defensiveness. And as dreams fail to materialise, so hopes, aspirations, for life dwindle, and people dwindle, become more entrenched and inflexible with the passing years. And the young look on and think in horror: "Is this in store for me? If it is, we must change the world *now*, so that I do not end up like that."

And because of all this identification of self with dream, and the resulting defensiveness, tension, bitterness, painful dwindling rigidity and loss of freedom, the very *activity* of light-hearted bubble blowing and bursting becomes impossible. The whole area of aims, dreams, ideals, utopias, heavens in the sky or here on earth, becomes set about with too many ferocious terrors. All too seldom could one announce at a dinner party say, as a delightful joke to be explored: I have a new plan to save humanity. Or: I have just thought up a new philosophy of life. Quickly, the good hostess would steer the conversation onto safer lines, to avoid an inevitable clash of frenzied and idiotic passions. To announce a new plan for the salvation of humanity was to assume the authority of prophet, the status of guru, the desire to be world dictator: in short, one announced that one was mad. In order to play at bubble blowing in public one had to be, it seemed, either a prophet or a madman: the idea that one could simply be *oneself* did not seem to occur to people. Bubble blowing was made out to be *so* important that only very special people could conceivably take upon themselves the responsibility of doing it for the rest of us: thus, if one did it oneself, openly, as a game, one was being all at once offensively flippant, impossibly arrogant, and ludicrously insane. No wonder light hearted bubble blowing failed to materialise. Just because it has been made into something so serious, so solemn, it has become something impossible to do straightforwardly and easily – and as a result it has lost much of its value, much of its potential for delight and use. Our philosopher suspects that bubble blowing has been invested with such absurd awe-inspiring solemnity for the following reason: people hunger to be told what to do, how to think, what to be, so that they can be relieved of making their own decisions, taking responsibility for their own lives: and at the same time people are terrified of this hunger, for of course it lays one open to being exploited by someone else, becoming the slave of another's will, at the expense of one's own best interests, and one's own freedom. Thus, only someone with the highest possible credentials, the most perfect disinterestedness, can be listened to: an ordinary person in bubble blowing would surreptitiously be seeking to enslave the human race. Do not the great philosophers – and especially Plato – seek precisely to do this? Is not the central, traditional problem of bubble blowing – one might almost say the central, traditional problem of knowledge – simply: How does one acquire the appropriate status, the necessary authority, to be in a position to settle these questions for everyone else? Typical answers: meditation, prayer, fasting, hearing the voice of God, inspiration, seeing the form of the good, having behind one the power, the authority of reason (or of science). And this absurd problem, and these absurd "solutions", are simply the product of the desire, and the fear, of so many people to be told what to think, what to do – and the desire of a few to tell the others what to think, what to do.

The same desperate ambivalence, it seems to our philosopher, haunts many people in relationships of love. People in love may long to lay down the felt burden of their responsibility for themselves, their "loneliness"; and at the same time they are terrified of the enslavement, the loss of liberty, that this implies.

Our philosopher knows all of this foolishness just because he is human too (in his own puppet fashion) and shares in it all himself. But he also believes in the delight, and the value of bubble blowing. In fact, for him, light-hearted bubble blowing and bubble bursting constitutes the heart of reason. The consideration which has led him to this view is extremely simple and can be put like this: If we choose bad aims, bad dreams, bad ideals, as we are quite likely to do, then the more *rationally* we pursue these bad aims the worse off we shall be, the further we shall be from realising good aims, that which we really want. In short, once we have misconceived or misinterpreted what it is we really want, reason becomes a menace, something which takes us *away* from what we want. Hence, true rationality must involve making every effort to ensure one has chosen good aims. And how does one do that? Well, it need be no effort: all that is required is light-hearted, enjoyable bubble blowing and bubble bursting. In that way, one supplies oneself with a rich store of vividly imagined and closely scrutinised bubbles, dreams, aims, ideals from which one may hope to make a good choice. Science ought itself to practice light-hearted bubble blowing, if it is to be truly rational. For real scientific rigour, there ought to be, at the centre of the scientific enterprise, a kind of riot of wild, delighted, free exploration of desirable possibilities. Academic enquiry in general ought to practice bubble blowing. Education should encourage and promote bubble blowing and bubble bursting. Gradually it would begin to spread, to catch on, to become an enjoyable, delightful pastime, a joke, an instinctive activity. People would be released from the single desperate prisons that haunt and imprison them. Unguessed riches would appear in *life*. Self confidence would grow. People's lives would flower. Society would flower. Everything that we all so passionately and so despairingly desire would come to pass. Reason, joy, freedom and laughter would walk hand in hand together. People would live in delighted harmony with one another, prizing difference, not fearful of it. And all this that we have dreamed of, and despaired of ever realising, is actually a realistic possibility, a practical proposition, something that we really can make happen. The essential step is to instigate light-hearted bubble blowing and bubble bursting. That is our philosopher's bubble!

And what happens? Just because it all means so much to him, off he goes, thundering away at the need for bubble blowing, invoking reason as his authority, especially when opposed, hurling thunderous accusations at scientists and academics in particular who, in his eyes, betray reason so profoundly even while they invoke "reason" to debar bubble blowing as a rational activity! In short, he identifies himself with his bubble. His *practice* is utterly at odds with what he *preaches*. Or rather he preaches instead of *practicing*! How absurd!

37

My gentle authorial puff of liberating good humour, blown into our poor philosopher's troubled mind and heart, has had the desired effect! It has stirred up these thoughts, these feelings, in him (they are *his* thoughts although I must confess, simply in order to alert the reader to a possible bias of presentation, that I, personally, find considerable sympathy with our philosopher's reflections). Our philosopher feels restored to good humour, recalled to himself at his own absurdity.

And off he goes to pay a visit to his friend the scientist to see if he can do a bit of enjoyable, scientific bubble blowing, without trauma, without anger, without sanctimoniousness, patronage or emotional dishonesty, but just for the fun of it.

Timidly he knocks on the scientist's door. It is three or four days after the last meeting.

SCIENTIST: Come in.

(Philosopher pokes his face round the door, grinning mischievously and a little uncertainly.)

SCIENTIST: Oh, it's you. The man who wants to put science on the psychoanalyst's couch. Come in!

PHILOSOPHER: May I? I was not quite sure whether we were still friends after our last explosion.

SCIENTIST: What does a bit of anger matter? Feelings aren't blows. I don't mind an emotional rough and tumble every now and again, as long as it doesn't get too serious.

PHILOSOPHER: (Beaming all over his face) Oh! I am so glad to hear you say that. I have these left wing friends who believe that ideally everything ought to be sweetness and light *all* the time: and I love to work myself up into a passion – at least on occasions – hurl furious words around with blithe indiscrimination, and then just burst into laughter at the absurdity of it all! And with them I can't do it at all. They just get cold, hard, angry: they begin to *hate* me for disrupting their inner peace and serenity. And that wasn't my intention at all. What is serenity *worth*, I ask myself, if it is achieved at the price of never being able to *enjoy* passion, even rage? Why not fling out our feelings sometimes? Among friends?

It doesn't mean you have to throw to the wind all consideration for others: if it is going wrong, then of course, one should stop. As you said, feelings aren't blows. And besides, behind our pent-up anger may be our pent-up creativity, our strong, free impulse to act. If we are afraid of our feelings we are afraid of *ourselves*.

SCIENTIST: (Dryly) Philosophy must be a peculiar subject.

PHILOSOPHER: You mean because it gets you to spout away all the time?

SCIENTIST: Something like that.

PHILOSOPHER: You're right! It is a peculiar subject. Unless it's *me*. Most probably *both* – and the combination is too much (and he thinks: "I must get this compulsion to deliver these personal, emotional mini-lectures all over the place under control. Just because I care so passionately about these issues, it doesn't mean everyone else wants to be deluged with my long-winded enthusiasms".)

SCIENTIST: Would you like a cup of coffee?

PHILOSOPHER: Yes please.

SCIENTIST: (After a pause, carefully, as he gets the coffee) What I don't quite understand about what you were saying last time is why you think your ideas have a special relevance for *science*. As far as I could gather, from what you were saying, the main culprit, in your view, is the philosophy of science, not science itself. That idea immediately gains my sympathy. I am inclined to regard academic philosophy as, intellectually, a pretty rubbishy subject. If one compares science with philosophy, then the contrast is very striking. Science meets with extraordinary *progressive* success. Philosophy never seems to get anywhere. Everyone seems to disagree with everyone else. No one seems to have any intellectual *respect* for anyone else. Philosophers seem, as a breed, to suffer from a quite extraordinary intellectual arrogance, even megalomania, which, given the facts, does not exactly seem to be called for. And what philosophers actually produce seems to fall entirely into two categories. It is either stirring, immensely ambitious *rubbish* – blueprint after blueprint for the cosmos, no less – (I am thinking here of the "great" philosophers: Plato, Descartes, Spinoza, Leibniz, Kant, Hegel, Wittgenstein, perhaps), or it seems to be (to the eyes of an ignorant outsider) so much sound, careful, analytical, technical, sterile *rubbish* – immense intellectual skill, balancing on a pinpoint, establishing *nothing*. (Here I think of the odd bit of contemporary academic philosophy I have looked at.) In either case, I am afraid, philosophy seems to me to be pretty rubbishy.

You agree! So why not address your remarks to your colleagues, who do perhaps need to pull their socks up, rather than to us scientists who, by and large, are not doing too badly, especially when one bears the contrast in mind. Put your *own* house in order, as a philosopher: then come to us with your new approach, your new ideas, and I am sure you will find us listening with interest and sympathy! (The scientist smiles warmly: he has spooned out these remarks as carefully as the Nescafe he has spooned into the two cups on his desk: and what he wanted to achieve he *has* achieved. I must try not to over excite this strange, excitable fellow, he thinks.)

PHILOSOPHER: In principle, of course, you are absolutely right. But, oh, I cannot tell you how utterly disheartening, wearying and soul destroying it would be to attempt to get my simple points across to my academic colleagues. It would take forty, fifty years of unremitting, soul destroying labour, a long uphill struggle against indifference and hostility, and even then it would in all likelihood be a complete waste of effort. Life is too short, and too precious, to be wasted like that. And what I have discovered is, I believe, too valuable to be buried in such a foolish fashion. For if I were to address myself in the first instance to my academic colleagues then merely in order to be *heard*, merely in order to be *published*, I would have to dress my simple, simple points up in such an incredibly intricate, technical form that no one else, apart from academic philosophers, could possibly hope to understand what was going on. Even I would probably lose track of what I was trying to say myself. (Something like that has already happened to me.) And in the end the whole effort would probably be wasted because the academic philosophers would still refuse to hear, and would even be unable to hear, just because the central simple points would be lost amongst all the technical disputation. Academic philosophy is, I am afraid, a lost cause. The communication lines have become too technical, too stereotyped. Long ago I gave it up. The only explanation that can be given for what most academic philosophers do at present is historical: they have been led into their present *cul-de-sac* by a persistent, long term failure to understand the true nature of their fundamental problems. Academic philosophers are so frightened that they don't have a subject! They have this absurd idea that other disciplines have been progressively stealing away with bits of their precious substance – physics, cosmology, logic, psychology, linguistics, political science, sociology. Desperately they cling to anything they can – any odd little puzzle – in an attempt to ensure that they have a subject that really does exist. They don't really *want* major philosophical problems solved, because they fear that then they would be out of an occupation. So they make their problems more and more difficult, more and more technical, and are delighted with the progress that they are making! At last, philosophy is being established on a firm foundation! The awful fear that the subject simply does not exist can be finally laid to rest.

And if only they would forget about their precious "philosophy" and turn their attention to *reality*, to life, to the great big, extraordinary, frightening, terrible, beautiful world, there outside your window, and here in your study. If only they would take as their central task, their central aim: to help improve our knowledge, our understanding, our appreciation, our enjoyment of the world and each other. To encourage open, easy, untraumatic, exchange of ideas about the world, about life between people. To help develop some ideas which might be of real use and value in life; help us to solve our enormous political, social, economic, moral problems; help us to work towards a better world. It is so simple. So obvious. So utterly childish. And yet – if I leave my personal friends and students on one side – there is only one contemporary academic philosopher of note who I am aware of who really pursues philosophy in such a spirit. I have in mind Karl Popper. He knows that the real business of philosophy is to help improve our knowledge, our understanding, our appreciation of the world and of ourselves. Not for a moment does he think, as most academic philosophers seem to: my concern cannot be with the *world*, for the natural sciences study the world; my concern cannot be with life, with society, for the social sciences study life, society; so my concern must lie elsewhere (but where?) Popper does not think like that because, for him, all this dividing up of areas of study into academic disciplines is ridiculous. Reality does not come to us, neatly packaged into "physics", "sociology", "inorganic chemistry", and so on. Often these distinctions between disciplines are only of academic administrative convenience, historical accident. It does not matter what academic label you carry. The chief thing is to improve our knowledge, our understanding, our appreciation of the world and ourselves. There are highly technical specialised problems. And there are broader, more sweeping, often more fundamental, less technical problems. A "philosopher" with a sympathetic interest in science, but not necessarily having a detailed, specialised knowledge of all the sciences in all areas (who could conceivably have such a thing?) may well profitably tackle the second broader category of problems, which arise out of attempts to understand the world. Someone ought to be tackling these important broad problems. By and large, scientists don't because they think: oh, that is too vague, too "philosophical". Philosophers don't because they think: our business is not to improve our knowledge and understanding of the world and ourselves, since that is what the natural and social sciences are doing, and we cannot very well intrude on their specialised territories. So no one tackles these problems (apart from the odd academic freak like Popper). The simple question: What does it all mean? What does it all mean to *me*, an individual person? never gets asked, or never gets discussed.

> Things fall apart; the centre cannot hold;
> Mere anarchy is loosed upon the world,
> The blood-dimmed tide is loosed, and everywhere
> The ceremony of innocence is drowned:
> The best lack all conviction, while the worst
> Are full of passionate intensity.[1]

Actually, now I come to think of it, the situation is even worse than this would suggest. Many academic philosophers believe that what they have to do is something called "conceptual analysis". It involves polishing up basic concepts, such as "knowledge", "truth", "good", "freedom", "mind", "person", and so on. At one time I thought that this activity was merely pointless, hollow. Gradually it began to dawn on me: the activity is actually both dishonest and counterproductive, something which actually *prevents* one from thinking. To have a nice polished set of concepts is, in effect, to have a view of things, a cosmology, a *Weltanschauung*, a philosophy of life. This is because theories about the world, about *life*, are invariably implicit in our concepts, however vague these theories may be. What philosophers *ought* to be doing is throwing open new possibilities, entertainingly indicating *Weltanschauung* that may not have occurred to people. The great thing is flexibility, mobility, creative richness. What conceptual analysis does is to present just one *Weltanschauung* (or a bit of one) as if it were a conceptual necessity, so that to think differently is to think meaninglessly. Quite literally, academic philosophers are trying to get the rest of us to stop thinking.

SCIENTIST: Can you give me an example of this?

PHILOSOPHER: Yes. A slightly old-fashioned example is Gilbert Ryle's *The Concept of Mind*.[2] In that book, Ryle advocates behaviourism, not as a straightforward (and obviously silly) theory about *us* (there is no inner experience, only behaviour) but as a *conceptual necessity*. Anything other than behaviourism is nonsense!

SCIENTIST: But how could anyone commit such an idiotic fallacy?

PHILOSOPHER: Put in four sentences, Ryle's argument goes like this. Is Cartesian dualism a meaningful theory? No, because if we analyse mental terms, consider their actual use, we discover that the relevant criteria are purely *behavioural*. Hence behaviourism is actually built into the meaning of our mental concepts. Hence Cartesian dualism is incoherent, meaningless.

Just how silly this argument is can be gathered from the following consideration: imagine a society divided up into *people* with Cartesian minds, and *robots* who *behave* just like people, but who do not have Cartesian minds (perhaps because their brains are made of transistors rather than neurons). Suppose further that the people believe correctly that the robots do not have minds. In such a world, Ryle's kind of "mindless" linguistic analysis would reveal that possession of a Cartesian mind was an essential condition for being a person.

SCIENTIST: The whole thing sounds incredible.

PHILOSOPHER: I assure you, it still goes on, in new dress, although on this quite grotesque point there are a few recent indications of some academic philosophers at last stirring themselves from their slumber.

SCIENTIST: It's lucky we don't have an academic philosopher with us today.

PHILOSOPHER: How right you are! There would be an almighty explosion.

SCIENTIST: At least you conceded that science is of value, even if a trifle neurotic. But philosophers seem to have nothing of value, if you are to be believed.

PHILOSOPHER: Well, of course, if I were speaking to a philosopher, I would arrange my remarks in a more tactful fashion. I would just say: the central concern of philosophy should be to encourage us – all of us – to share our ideas about life, about Nature, about our hopes and dreams, in a friendly, sympathetic, enjoyable way. And they would laugh; perhaps even agree, verbally. And who knows; there might be a little nudge in the right direction.

SCIENTIST: And is that how you are treating me now? With such patronising tact?

PHILOSOPHER: No, no, no! Certainly not at the moment, at any rate (they smile). And in any case, we have agreed: there is an immense difference. Academic science, everyone would agree, is unquestionably of importance in our modern world. Whether anyone other than academic philosophers could say the same for academic philosophy is another question.

But you can see the problem. The one thing that cannot be said is simply: The Emperor has no clothes! Only children are permitted to say such things: and they soon get the impulse knocked out of them. People have such a terror of experiencing themselves as naked, shorn of their precious clothing of beliefs, values, ideals. They put their identity, the meaning and significance of their lives, into their clothes; so that, not surprisingly, when someone comes along and says casually and convincingly: "Hm, pretty shoddy lot of patched up rags you have on here", the remark is experienced as a terrible threat, a challenge to the whole meaning and significance and value of that person's life. All the finery has been reduced, at a touch, to rags. From Emperor to beggar. But if one did not mind occasionally experiencing oneself as a naked human being, then this absurd identification of oneself with one's ideas and ideals would not be so remorseless. Instead of doing all one's thinking, feeling and acting through one's own clothes, through one's own cherished beliefs and ideals and as a result being able, only very slowly and painfully, to modify or to develop these beliefs and ideals, one could every now and again step right out of them and put on some quite different clothes – the clothes of a child of six, for example – and see how things looked and felt from that perspective.

SCIENTIST: It all sounds a bit mystical to me.

PHILOSOPHER: I don't think it is really. I believe it is just plain commonsense, something that every child knows and experiences instinctively, unthinkingly, but something that so many adults have entirely lost sight of. One of my grimmer fantasy pictures of the world, in fact, goes like this. Children (those who have not been put into invisible chains that is, at an early age) are full of liveliness, curiosity, wonder, instinctive passionate response to the world around them. The world is full of mystery, colour, brightness, charged with meaning and value – sometimes frightening, sometimes enrapturing. (Have you watched children discover things? Look, a bird! Look, the moon! They point, and are transfixed with unaware rapture, which we notice, envy and pretend not to notice.) But as children grow up, something frightening begins to happen. Children gradually become entangled in a fixed set of clothes – a straightjacket one should perhaps say, rather. Fixed aims, beliefs, values, ideals trap the naked person. The person is condemned to live out just that pattern of life, paralyzed, utterly immobilised, apart from the permitted actions, thoughts, feelings within the tight prison. The inner child is struck dumb with horror: but try as it can, there is no escape, unless through sudden abrupt conversion to a different straightjacket – a loose strap slips a little – or of course into madness. The child within despairs, and gradually dies, a dead, lost world thick with cobwebs within.

If one really wants to give oneself the horrors, one can even imagine that this is *neurological* in character. Perhaps some central controlling part of the brain gets locked into a fixed position, so that only a fixed way of life can be pursued. It may be that a child who does not keep flexibility alive during childhood, will have lost the capacity for ever, much as a child who has not learnt to speak a language by the age of twelve has for ever lost the capacity to speak.

Each generation of children try their best to wake us up. They leap around: they shout at the tops of their voices; we frown and tell them to shut up and be still. They badger us with questions: Why? Why? Why does it have to be like this? They see the grey world waiting. Most probably imagine it to be a world of extraordinary hidden mystery and beauty; they long to join it! A few are more suspicious. And then they begin to ask: What does it all mean? Why is one alive? What is there to live for? Suddenly, or slowly, they get their answer. Snap! The answer has swallowed them up.

SCIENTIST: But there are other explanations for what you are talking about. Adults find themselves forcibly shut in a treadmill not of their making. Others put their shoulder to the heavy wheel out of a sense of responsibility. Children can afford to be irresponsible.

PHILOSOPHER: You are right.

SCIENTIST: And in any case, things are not as bad as you make out.

PHILOSOPHER: Again, you are right. I was only blowing a gloomy bubble, for the hell of it, to be burst at the faintest pinprick of criticism.

SCIENTIST: A bubble?

PHILOSOPHER: Oh, that's just a technical term, a stitch in this forage of nonsense, my philosophy, in which I sometimes wrap myself up at night, to keep myself warm. Think of it as a fantasy, rather. An outline for a science fiction story. The evil aliens have been putting something nasty in the water again. Sleeping sickness virus, perhaps.

SCIENTIST: We seem to be wandering rather from the subject – which I take to be, for the moment: Why is academic philosophy so sterile? But before I ask you about that, two points worry me about what you have just been saying.

PHILOSOPHER: Yes?

SCIENTIST: You say academic philosophy is mostly pretty sterile. These dignified professors of philosophy, dressed in their fine academic robes, are really intellectual beggars.

PHILOSOPHER: Yes.

SCIENTIST: Well, perhaps it is the other way round! Have you thought of that? Perhaps *you* are the beggar, with delusions of being dressed as an Emperor!

PHILOSOPHER: Of course! All too possible.

SCIENTIST: How do you get out of that then?

PHILOSOPHER: By not taking it all too seriously. Beggar, Emperor, we're all human underneath. Life before thought!

SCIENTIST: Alright. But another thing worries me about what you have been saying. This stuff about childhood.

PHILOSOPHER: Yes?

SCIENTIST: Well, it seems to me that what you were really doing there was to pull an old trick. You were claiming for your side of the story the Authority of Childhood – an old idea of Romanticism. Childhood uncorrupted by the ways of adulthood. You have remained a child, and hence you can see what these eminent grey philosophers can no longer see. So you must be right!

PHILOSOPHER: (Admiringly) That is very sharp!

SCIENTIST: We have to be sharp in our profession.

PHILOSOPHER: I really do believe there is *something* in that stuff about childhood. But of course you are right. In context it could look as if I were making a very surreptitious, sneaky appeal to Authority – to a very appealing authority at that: the gentle, defenceless authority of childhood. I had no idea. It just slipped out. You *are* a cunning bastard!

SCIENTIST: And now, could you please stop prancing around all over the place, and answer in simple straight terms the following questions. You claim that academic philosophy is nothing but a waste of time. What I would like to know is this. Doesn't that judgement smack a little of intellectual arrogance? Why am I supposed to think you are any different from the rest? You are, after all, professionally, an academic philosopher. Third, in simple, clear terms, *why* do you think academic philosophy is in such a mess? What *ought* they to be doing? Why aren't they doing it?

PHILOSOPHER: Let me take your questions in turn. I can't help sounding intellectually arrogant. I certainly didn't mean to get myself in this position. I simply pursued my problems, my intellectual concerns and interests as honestly as I could, and found myself as a result in this socially awkward situation. As to being unlike my academic colleagues: of course I am not! You, yourself, said it: "Everyone disagrees with everyone else. No one seems to have any intellectual *respect* for anyone else. Philosophers suffer from intellectual megalomania". I am exactly like the rest in all these respects. But we are *all* more or less like this. We *all* have our personal philosophies which declare: A, B, C are worthwhile, important, meaningful human pursuits; X, Y, Z are stupid, meaningless, pointless pursuits. It's just that my personal philosophy says: academic philosophy, as it mostly goes on now, is mostly rubbish. In fact, on this one point, our two personal philosophies are in *agreement* – yours and mine. Leaving entirely on one side the long, sad, historical story, the basic trouble with academic philosophy today can, in my view, be pinpointed like this. Academic philosophy *ought* to be concerning itself with *life*, with philosophies of life, with encouraging and promoting the rational evolution of philosophies of life – and so the evolution, the flowering of *lives*, in the direction in which people themselves choose, thus enhancing individual freedom. Philosophy *ought* to be concerned, quite straightforwardly, with helping to promote individual liberty. (This concern may, of course, group itself around some particular human endeavour, such as education, science, mathematics, politics, art, literature, and so on, thus developing helpful ideas, possible clarifications of aims and methods, designed to help the pursuit of these endeavours.) All this is what academic philosophers *want* to do, and often try to do. However, in addition, they want to do philosophy in an authoritative fashion. They want to make a contribution to *knowledge*. They want to put philosophy on a sound academic footing. And this second aim or aspiration, combined with the first, creates an intolerable dilemma. For how can the *philosophy of life* be dealt with in an authoritative fashion, as if it were a department of *knowledge*? How could reason, intellect, an academic subject, deliver verdicts on high about how the rest of us should live? The very idea is an absurdity!

There are two ways in which one may attempt to avoid this absurdity.

First, one may hold onto the desire to have a respectable, authoritative academic subject, that constitutes a branch of knowledge, and abandon, as a result, all claims to be concerned directly with life, with problems of life, with philosophies of life. One seeks to talk around the subject: philosophy becomes, as it were, a meta discipline; the philosophy of philosophies of life. Ethics is itself value-neutral. Philosophy of science is itself quite distinct from science. Philosophy of art has nothing to do with furthering art, helping art to flourish; it is concerned only with problems of knowledge and understanding concerning the nature of art. Philosophy of politics ought to have nothing to do with helping to settle political issues. The philosopher looks down on the human comedy and tragedy from Olympian, uninvolved heights, and delivers up his philosophical analyses, concerned only with the philosophical understanding of life, not with life itself. All this inevitably leads, I believe, to a combination of intellectual dishonesty and sterility. It leads philosophers to concern themselves with the problems of *philosophy*, instead of the problems of *life*. The problems of philosophy become increasingly dissociated from the problems of life, so that the very idea that academic philosophy might have some kind of relevance to life becomes increasingly absurd.

The second option is to throw away altogether the idea of philosophy as an authoritative branch of knowledge – at least in anything like the conventional sense. A philosopher is no kind of expert, no kind of specialist, and has no authority whatever. If anything he is a professional jack-of-all-trades, a professional dilettante. A sensible – even potentially useful – aim for a philosopher to pursue can be conceived of in the following terms. Our philosopher should become sympathetically involved in those aspects of life that interest him, concern him personally: those aspects of life which he especially values. He should seek to enter sensitively and imaginatively into the relevant human pursuits. He should involve himself in these pursuits. And his fundamental *philosophical* concern should be simply to come up with a few *suggestions* – suggestions only – as to how aims, ideals, values might be improved, made more desirable or more realistic; and, in addition, his concern should be to suggest a few *methods* that might be of help in realising desirable aims. Philosophers should turn their back on philosophy, and the problems of philosophy, and concern themselves with life, and the problems of life. Or, at the very least, philosophical ideas need to be developed so as to make fruitful and helpful contact with life, rather than being utterly removed from life.

And philosophers, above all, should keep going a certain modesty. It is people who actually *do* things well, who are almost always the best equipped to say *how* to do these things well. Philosophers who can do nothing but philosophy (very badly) are the least equipped to go around telling others what to do. At the most, they should seek to set up a dialogue with the doers. "Is this it?" they should ask, "Are these your aims, your ideals, your methods? Is this how it feels? No? Then tell us! Tell us!" Philosophy is much too important to be left to academic philosophers. Perhaps academic philosophers should regard themselves as the careful, sensitive custodians of the philosophical ideas of the *doers*. But, of course, these very distinctions are a bit silly. We are all doers. Even academic philosophers have interests outside academic philosophy (their saving grace!).

SCIENTIST: You have changed your tune a little from last time, haven't you? Last time you were fiercely, arrogantly denouncing science as neurotic. Now you merely wish to make a modest suggestion, for *our* consideration.

PHILOSOPHER: Oh, only if you knew how it has been! Really, that is all I do want to do, make a *suggestion*. I *care* about science; I am involved; I believe in it, value it. And, understandably, I want to make my own contribution to this magnificent cooperative endeavour. What happens? I am rebuffed! My contribution does not fit the conventional rules of the game: of course not! It is precisely the suggestion that these conventional rules need perhaps a closer examination. That which I am criticising keeps my criticism at bay. I am told to talk to my colleagues, the academic philosophers. But it is *science* I care about; and more, *life*, my life, our common life, here on this planet of ours. Where does one speak if one has such a concern? Believe me, I am sorry about that outburst. That very definitely is *not* the way to proceed. It goes against everything that I am advocating.

SCIENTIST: I think I am at last beginning to get a glimmering of what you are on about. Your sympathetic interest in *science* has led you to dream up a *suggestion* – a suggestion only – as to how it might be possible to improve things a little. And you would like the scientific community simply to *consider* the suggestion. It is, of course, entirely up to the scientific community to *decide*. Probably your suggestion is impractical, not up to much, just because your personal experience of scientific research is rather limited. Nevertheless, you think your suggestion deserves to be considered. Is that it?

PHILOSOPHER: (Absolutely delighted) That's it, exactly. The only point that I would add is that in my view science is a community affair, something that ought ideally to concern us all. In my view, science can only be truly rigorous, rational, objective if it is the possession of the whole community, as it were, not the exclusive preserve of specialists. That is what I mean by 'person oriented science', Pygmy science. Thus, my concern is to address my suggestion to the community as a whole, and not just to scientists.

You can see the problem. According to my view of philosophy, philosophers ought primarily to be talking to non-philosophers, encouraging people to articulate their philosophies, in the straight-forward commonsense notion of 'philosophy' – that is, their aims, ideals, aspirations. But for so long the academic philosophers have been producing such irrelevant rubbish that everyone else has got fed up. Who wants to talk with a philosopher these days? Thus, real philosophy, community philosophy as it were, as opposed to the professional philosophy, can scarcely get off the ground. Interconnections between things remain unnoticed, or unarticulated. Look at the difficulties we have had in simply getting round to talk with each other with a small measure of mutual sympathy!

SCIENTIST: Look, I have had enough, I think, for today. How about returning to the real topic of our discussion on some other occasion? I have some *work* to do!

PHILOSOPHER: Alright.

CHAPTER FIVE

OUTLINE OF THE WHOLE ARGUMENT

(This time the knock comes on the Philosopher's door, and the knuckle belongs to the Scientist.)

PHILOSPHER: Come in.

(The Scientist appears, grinning a bit sheepishly.)

PHILOSOPHER: (Delighted) Come in! How very nice to see you.

SCIENTIST: I've come to finish off our discussion.

PHILOSOPHER: Excellent.

SCIENTIST: I feel that up till now we have really just been skirmishing, fencing around the main subject. I want to hear what your suggestion is, in plain and simple language. Not that I really believe that there can be anything very much in it, let me add. I don't want to raise your hopes unduly. But I confess: I am curious to hear what you have to say.

PHILOSOPHER: I am overwhelmed!

SCIENTIST: OK, cut the cackle.

PHILOSOPHER: I *mean* it.

SCIENTIST: O.K. Now as I understand it, your position is this. Basically science is a sound intellectual enterprise. It makes genuine progress. It is of real value to humanity. It has a fundamental intellectual honesty, intellectual integrity. It achieves genuine knowledge and understanding concerning natural and biological phenomena.

PHILOSOPHER: (In a great burst of enthusiasm) But it goes so, so, so much further than that. You said it when we first met. Modern science has helped to transform our world infinitely for the better if one looks back to Medieval times say, or even Victorian times. It is not just all our comforts and luxuries, our modern methods of transport and communications, our radios, televisions, hi-fi sets, which make us so careless of all the information we possess, the great music we can hear whenever we please, at the touch of a button. Above all, there is modern medicine, modern methods of hygiene, made possible by the scientific understanding of disease. A life of a child, saved by some simple operation: we in the West have, perhaps, lost imaginative contact with the casual brutality of life here in earlier times. There is an immense fund of selfless dedication and compassion behind so much of scientific research. And it is all the greater for being so practical, so effective, often so unimaginably beneficial in its results, and yet so calm, matter-of-fact, undemonstrative.

And look what science has done for our understanding and appreciation of the world around us. I do not hold much truck for the social sciences. But the natural and biological sciences are different. Wherever one turns, scientists have been there, questioning, gazing, seeking, with a kind of grave loving concern. In imagination we can peer into the vast depths of space and time. Our knowledge extends far beyond our earth, far beyond our solar system, beyond even our galaxy, to the myriads of galaxies that go to make up the known universe. And we have discovered strange new objects in the sky, neutron stars, quasars, and even perhaps, who knows?, black holes. We have even detected traces of the original cosmic convulsion from which, as far as we know, everything began. We have delved into the fundamental laws of nature, and have come up with extraordinary discoveries about the ultimate structure of the universe: gravitation as curvature of space-time; fundamental particles mysteriously sharing wave-like and particle-like properties; mass as a form of energy; the very stuff of the universe, space, time, energy, mysteriously interwoven in some as yet little understood way. The long evolution of the sun, the solar system, our planet, lies before us in essentials for us to contemplate. The extraordinary story of the development and evolution of life on this planet is gradually being unravelled. Here we are now, the unintended consequences of all those courageous, bright-eyed animals, struggling for life in a world which *they* did not understand – or not as well as we understand it. It is a miracle, a tale that grips the heart with wonder. Pick up a book on zoology or botany, and one can learn at one's ease of the incredible diversity and complexity of life forms that have evolved on earth. The story of our own evolution and history has been told again and again, with a wealth of brilliant and accurate detail, from many different viewpoints. The ways of life, ideas, values, customs, of cultures and societies profoundly different from our own are available for our sympathetic understanding and appreciation – so that we may learn from them. Our knowledge and understanding of the chemical and molecular workings of life has been enormously enhanced in recent years. As *you* know, we now understand in principle much of the molecular basis of inheritance. A great deal is known of the extraordinarily rich, complex and beautiful systems of chemical and molecular controls responsible for the maintenance and preservation of our living bodies. And we even know something of the almost unbelievable complexity of our brains – that which gives us thought, feeling, consciousness, self-awareness, our identity as persons. It really is the simple, plain truth: modern science is the outcome of a kind of profound reverence for the world in all its multifarious and extraordinary aspects. With immense patience, care, attention to detail, scrupulous honesty, objectivity and self-criticism, science seeks to know and to understand. One might almost say that science is the flowering of our finest loving attention for our world in all its aspects, from the first few milliseconds of the existence of the cosmos, to the detailed structure of a fly or virus. And the rich store that is the outcome of all this seeking and searching is in principle open for us to plunder, to use as food for our imagination. If the human point of science, from a cultural perspective, is to put us in touch with reality, extend and deepen our experience of the world, reveal to us that which is significant and hidden, then surely modern science has, in essence, magnificently achieved this aim. Looking back to the ideas of earlier times, we can surely only declare that the vision of the world then available to people was in comparison extraordinarily restricted and distorted by inadequate knowledge and understanding. Are we not profoundly enriched in our lives by this beautiful, far-reaching, imaginative, and immensely detailed, carefully checked vision of our world that has been assembled with such care, with such concern, by modern science?

And it is not even the case that all this knowledge and understanding that we have of the world has killed off the mystery of things. For the truth is: we do not *know*. As Popper has emphasized, with such clarity and passion, all our knowledge is but a tissue of beautiful and amazingly successful guesses. We are not, as it were, in a position of power or authority over the world, capable of predicting, with absolute assurance, what Nature will do, in such and such circumstances. At any moment, and in any way, for all we can know for certain, Nature may surprise us utterly, by failing to live up to our expectations and predictions. Our knowledge is but careful, attentive, loving, trusting speculation – a magnificent chorus of Pygmy songs to our beloved cosmos.

In fact, if our basic guesses about the nature of the universe are not wholly astray, then we *know* that our present knowledge and understanding is limited and imperfect in all kinds of ways. For many of our ideas and theories fail to fit together, to cohere. There are all kinds of gaps, problems, enigmas, mysteries. If we conjecture with Einstein, as I think we should, that a beautiful unified mathematical pattern runs through all natural phenomena, then we must confess that we seem to be a long way from discovering this pattern. Our present-day fundamental physical theories seem only to provide us with distorted, inconsistent glimpses of restricted aspects of this conjectured unified pattern. Our two fundamental physical theories – quantum theory and general relativity – refuse to fit together in any harmonious way at all. I would say, along with Einstein, that quantum theory itself at present suffers from a basic inadequacy, in that it is not a fully micro-realistic theory, not a theory about how fundamental physical particles interact with each other, but only a theory about how these entities interact with macroscopic measuring instruments.[1] We really believe that macro phenomena are made up exclusively of micro phenomena: and yet quantum theory does not accord with this simple belief. To put the thing in slightly different terms, we scarcely have a glimmering of an idea for a unified theory of the four kinds of forces or interactions that there are to be found in the world – strong and weak nuclear forces, electro-magnetism and gravitation: in the end, we do not really know what sort of basic stuff the world is made out of – such is our ignorance! We understand very little of embryology, and of the workings of the brain – how neurological processes are related to consciousness, memory, perception, emotion. How life began on earth is still largely a mystery. Whether life exists on other planets, and especially whether *people* exist on other planets, is completely unknown to us. We seem to be incapable of understanding physical processes capable of producing the immense amount of energy apparently being emitted by quasars. Here are just a few areas of fundamental ignorance that we more or less *know* we have: for all we know, there may well be highly significant aspects of the world about which we are ignorant that we are ignorant!

Well might one say, with Einstein: "… all our science, measured against reality, is primitive and childlike – and yet it is the most precious thing we have." [2]

It's not just that science puts us in touch with reality, extends the range and depth of our experience of the world. Science puts us in touch with mystery, with that which we do not know or understand – *real* mystery, not conjuring tricks. Isn't that close to one of the supreme virtues of science – its capacity to provoke wonder, startle us out of our familiar ways of thinking and seeing, provoke in us a lively curiosity about the little known and little understood? And is there not, above all, the *spirit* in which science is pursued? Daring, beautiful speculation controlled by sceptical, critical attention to detail, to those awkward, exasperating and apparently trivial recalcitrant facts which do not quite fit in with the original sweeping idea. There is such wild imagining in science, such courageous envisaging of possibilities; and there is this care, this patience, this sustained capacity to doubt. It is so magnificent – the responsibility, the honesty, the spirit of cooperation, the practicality, the underlying *gravity* of concern. Scientists, themselves, do not effuse about all this like me because they do not think it necessary: they want to keep things calm, practical, realistic. There are passages in Bach when the music seems to come from some unimaginable depth of calm, grave concern for the whole human race, for the whole of reality. That is what I see in science. When I think ..... I'm sorry, I'm sorry, I am being stupid.

(What on earth has happened? Our Philosopher is weeping! The Scientist looks away feeling distinctly uncomfortable before this display of emotion)

PHILOSOPHER: Hm! (Wiping his eyes) Let *me* get *you* some coffee.

SCIENTIST: Thanks.

(An awkward pause)

SCIENTIST: You know, if I didn't know you better, I would say you were up to your rhetorical tricks again.

PHILOSOPHER: How do you mean?

SCIENTIST: Well, manoeuvring yourself into position so that you can all the more effectively deliver your punch line. Convincing us that your heart is in the right place, so that your criticisms, your suggestions go home all the more effectively.

PHILOSOPHER: (Laughs; his feelings seem to come lightly.) You are probably quite right! That's what I admire about science at its best. The sustained scepticism. You *scientists* don't madden yourself with words. (And then he adds, more seriously) Whatever my *motivation* may have been for giving that spiel just now, on this particular occasion – even if a slight deviousness of motivation was present – please believe me: I really do believe it all, feel it all, passionately. I really do believe that science is a profoundly beautiful and valuable human creation.

SCIENTIST: But!

PHILOSOPHER: Yes, there is a 'But'.

SCIENTIST: You have the audacity, the nerve to suggest that there is at present something seriously wrong with 'this magnificent cooperative endeavour of science', as you would probably call it!

58

PHILOSOPHER: (Grins wickedly) Yes, I do. (And then he adds) But please take it in the following spirit. In his delightful book *Prelude to Mathematics*, W.W. Sawyer, in talking about the qualities of a mathematician, has this to say: "Mental venturesomeness is characteristic of all mathematicians ….. For example, if you are teaching geometry to a class of boys nine or ten years old, and you tell them that no one has ever trisected an angle by means of ruler and compasses alone, you will find that one or two boys will stay behind afterwards and attempt to find a solution. The fact that in two thousand years no one has solved this problem does not prevent them feeling that they might get it out during the dinner hour. This is not exactly a humble attitude, but neither does it necessarily indicate conceit. It is simply the readiness to respond to any challenge." [3]

That is the spirit in which my suggestion is put forward. It is a bold conjecture, seeking admission for *consideration* only, by the scientific community. It is put forward entirely in the spirit of science: daring speculation, controlled by criticism.

SCIENTIST: Very well. And your suggestion I take it is that science has fallen foul of a bad *philosophy* of science which –

PHILOSOPHER: Exactly. This truly marvellous human creation, science, has been wrapped up in a quite grotesquely inadequate philosophy of science – a piece of tatty old newspaper. And as a result, the truly wonderful thing that science really is, is hidden from most people's eyes. Those creatively involved with science know all too well just how magnificent science is. But outsiders see only science wrapped up in tatty old newspaper. For them, science is cold, threatening, grim, unpleasant, unintelligible, dogmatic, unchallengeable, dangerous – altogether either *boring* or *obnoxious*. Scientists try to say: "No, no, no, science is not like that at all. It is quivering with life, with vitality, with care, with sensitivity, with imagination, with wonder. It is beautiful and valuable." But the bits of tatty old newspaper get into their mouths. It all comes out garbled. "Science is concerned solely with value-neutral *fact*. On the intellectual level, science is completely impersonal, unemotional, indifferent to beauty, to suffering, to the needs, frustrations, problems, desires, aspirations, ideals of *people*.

It just improves our knowledge of objective value-neutral fact, in a completely impersonal – almost inhuman – way." The passionate human heart of science has disappeared from view. Only something terrifyingly cold and impersonal seems to be left. And scientists actually hold that scientific, intellectual integrity *demands* that one adopts this kind of cold, inhuman attitude – so gripped are they by the deplorable ideal for rigour of standard empiricism. No wonder non-scientists say: "Let them get on with it then. I do not wish to have anything to do with it." The majesty, the beauty, has been lost.

SCIENTIST: Please. You really must try to control yourself. I do not have much time, I am afraid. Can we keep the discussion simple, without these emotional effusions?

PHILOSOPHER: I apologize.

SCIENTIST: Is it alright if I give a simple outline of your basic points, your suggestion, as I understand it, so that you can tell me whether I have got hold of the basic idea?

PHILOSOPHER: Yes. Excellent.

SCIENTIST: Your claim is that science has become gripped, as a result of a kind of intellectual carelessness on behalf of the scientific community, by a philosophy of science, an ideal for scientific rigour and integrity, which you call *standard empiricism*.

PHILOSOPHER: Yes.

SCIENTIST: And your claim is that standard empiricism exercises a profound; far-reaching influence over science, over the whole way in which science proceeds. It influences the intellectual domain of science, and the institutional, educational, cultural, social, moral domains of science. It influences the way in which scientists *think* about science, *do* science to some extent, speak about science. It influences criteria for publication of results, criteria for acceptance of results. The very *content* of scientific knowledge has come to be influenced by the widespread acceptance of standard empiricism. Scientific, intellectual *values* are influenced by standard empiricism. And above all the widespread acceptance of standard empiricism – and the resulting attempt to make science conform to standard empiricism – has exercised a profound influence over the whole way in which science is related to *life*, to *people*, to *feelings* and *desires*, to people's *problems, needs, values, ideals, aspirations*. Acceptance of standard empiricism affects profoundly the whole way in which science is related to society, in short.

PHILOSOPHER: Yes.

SCIENTIST: And you claim further that this profound, far-reaching influence of standard empiricism over science, and over the way in which science is related to people, has a seriously detrimental effect in all these areas. The intellectual progress of science is impeded. Bad educational, cultural, social, moral practices of science result. Scientists are led to think about science and, to some extent, to *do* science, in a bad way. Bad criteria for publication of results, and for acceptance of results, exist in science at present as a result of this influence of standard empiricism. Even the *content* of scientific knowledge is adversely affected. Our very view of the world is adversely affected by the widespread acceptance of standard empiricism. Scientific, intellectual *values* suffer as a result of the adverse influence of standard empiricism. But of far greater importance than any of this, in your view, are the appalling, bad effects that widespread unthinking acceptance of standard empiricism has for the relationship between science and people. It is because science is wrapped up in this bad philosophy of science of standard empiricism that people are unable to *see*, to *experience*, to *know*, the extraordinary, beautiful, majestic thing that science really is. The human *use* of science suffers. And as a result *we* suffer. Our relationships between ourselves and the world, and between ourselves and ourselves, are adversely affected, just because science ought, ideally, to be understood and pursued as an aspect of these relationships, and precisely this is forbidden by standard empiricism – is indeed almost incomprehensible given a standard empiricist conception of science. Ordinary people cannot use science profitably and painlessly to put themselves personally into touch with the cosmos, and into touch with other people. Individual people cannot *use* science to enhance their relationships with the world and other people, develop more knowledgeable, understanding, appreciative, harmonious, fruitful relationships. The widespread attempt to keep science in the straightjacket of standard empiricism – as you call it – in effect has the consequence of putting us *all* in straightjackets. The flowering of our lives is curtailed.

PHILOSOPHER: Yes.

SCIENTIST: And, according to you, all these bad consequences of the *attempt* to make science conform to standard empiricism flow from the simple fact that standard empiricism represents a quite grotesquely *unrigorous* philosophy and ideal for science. Scientists eagerly try to make science conform to standard empiricism from the very best of motives: to keep science rigorous, objective, rational, an authentic road to factual knowledge. And the tragic irony of this is that it is all the other way round: in seeking to make science conform to standard empiricism, scientists are actually *undermining* the rigour and objectivity of science. And all the bad consequences of standard empiricism flow from this simple point.

PHILOSOPHER: Yes.

SCIENTIST: We need a new, much more truly *rigorous* ideal for science – aim-oriented empiricism, humane aim-oriented empiricism, aim-oriented rationalism, person-centred science, I am not quite sure which of these it is you are putting forward.

PHILOSOPHER: They constitute a progressive tightening up of rigour.

SCIENTIST: Yes, I see. Your truly rigorous ideal for science is what you call the ideal of 'person-centred science.' And according to you, if this fiercely rigorous ideal for science were taken seriously, and put into practice, then all the intellectual and personal defects of science, as it exists at present, trapped within standard empiricism, would begin to disappear. The truly beautiful and magnificent thing that science really is would no longer be buried from most people's view. It would shine forth, for everyone to experience and know. As a result of developing science in accordance with the more rigorous ideal of person-centred science, the intellectual progress of science would accelerate. The educational, cultural, social, moral aspects of science would flourish. Scientists' understanding of science would be improved; their scientific activities would be helped, criteria for publication of results, and for acceptance of results, would improve. The very *content* of scientific knowledge would be subtly transformed for the better. Our knowledge, our understanding, our appreciation of the universe would be enhanced. Scientific intellectual *values* would be improved.

In your view, however, of far greater importance than any of this is the possibility of there developing quite different relationships between science and *people*. People could begin to *use* science in order to enhance their relationships with the cosmos, and with each other. Gradually, we could all begin to use science, in a personal way, in order to develop more knowledgeable understanding, appreciative, harmonious relationships between ourselves, and between ourselves and the cosmos. Despite the continued existence of specialized, technical aspects of science, the most precious *heart* of science could become generally available, childishly accessible. Indeed the heart of science could be said to be – er – the human heart. It would not just be science that would be transformed by putting into practice this more rigorous ideal of 'person-centred' science. *We* could be transformed – in ways which we, personally, desire and value. Person-centred science would be science experienced and practiced as our version of the Pygmies' song. Our hearts, our feelings, could come to be sensitively in touch with the objective beauty of the world. We could come to have what you would call, I suppose, objective feelings, feelings that – in your view – inform us of the objective nature of aspects of the world. Through our scientific theories, imbued with feeling, we would come to feel for objective reality. And our technology would be – um – the music of our best personal needs and desires delicately embedded in the music of Nature's laws, our songs "awakening" Nature, to help her help us. Theory and technology would be two closely inter-related aspects of improving our relationships with Nature. Gradually we could come to love and trust our world, and each other, in something like the way Turnbull says the Pygmies do. Science would help us to feel good about our world: it would (1) help to put us into touch with that about which we can feel good, (2) help us to create a world about which we can feel good and, (3) help us to discover within ourselves the capacity to have such good feelings.

There. I have done my best. Have I got the basic idea more or less straight?

(During this recital, the Scientist has been almost visibly struggling to get out words that he hopes will please the Philosopher – words that taste so peculiar in his own mouth.)

PHILOSOPHER: I am absolutely overwhelmed.

SCIENTIST: But you mustn't run away with the idea that I *agree*. I may *understand*: but that does not mean I *agree*.

PHILOSOPHER: It's just that which impresses me so. How rare it is for someone to go to the heart of someone else's ideas – even to the point of sympathetically improving on, and clarifying the expression of those ideas – and yet all the time quite fundamentally *rejecting* the ideas. If only that were more common: we would not have our present problems of communication at least.

SCIENTIST: Flattery won't get you anywhere. Oh, I've just remembered two additional points. One: according to you, reason and desire go hand in hand together: reason helps us to discover and attain that which is desirably desirable; if it doesn't, then there's something wrong with our concept of reason or with the way we are using it!

PHILOSOPHER: Yes.

SCIENTIST: Two: academic philosophers of science spend their time *attempting* to prop up the hopelessly inadequate philosophy of standard empiricism. They fail miserably, producing nothing but sterile rubbish, which serves only to convince scientists that philosophy of science is a waste of time. Thus, academic philosophers of science are a further part of the overall intellectual conspiracy.

PHILOSOPHER: Yes.

SCIENTIST: And now can I ask you one simple, direct question?

PHILOSOPHER: Yes.

SCIENTIST: In simple, brief terms, can you outline your *argument* which, you think, establishes that person-centred science represents a more rigorous ideal for science than standard empiricism? If I can have an outline of your whole argument before me, then perhaps I will be able to get a more sympathetic understanding of your more specific, more detailed remarks.

PHILOSOPHER: Yes, of course. The argument is, I believe, in essence childishly simple and obvious. It can be outlined like this.

We begin with *standard empiricism* which asserts: "The basic aim of science is to improve knowledge of value-neutral factual truth. This is achieved by choosing theories solely with respect to experimental and observational success, in a completely impartial fashion. Science only retains its rigour, its intellectual objectivity, its intellectual integrity, its entitlement to the claim to deliver authentic factual knowledge insofar as the *intellectual* domain of science is ruthlessly dissociated from the domain of personal, social needs, desires, aspirations, problems, feelings, values."

SCIENTIST: Yes; I've got that.

PHILOSOPHER: As an ideal of rigour for science, however, *standard empiricism* is a wash-out, for the simple reason that it grotesquely *misrepresents* the basic *aims* of science, the aims that science in fact pursues and the aims that science ought to pursue. Science does not seek to improve our knowledge of truth *as such*; rather science seeks to improve our knowledge of *valuable truth*, truth that is in some way or other important, significant, beautiful, interesting, useful.

SCIENTIST: But –

PHILOSOPHER: The moment we recognise this simple point, it becomes clear that the fundamental aims of science must remain permanently and profoundly problematic. What is there to discover that is valuable? What is valuable? Valuable for *whom*? It is of fundamental importance for science that we make the best possible choice of aims for science; and it is almost inevitable that we will fail to make the best possible choice of aims for science. Here, above all, then, we need to proceed intelligently, critically, imaginatively, wisely. If science is to be truly rigorous and objective, it is essential that scientists and non-scientists alike articulate, explore, develop and scrutinize possible and actual aims for science (cooperatively and delightedly engaging in scientific bubble blowing and bubble bursting, in fact), so that we may provide ourselves with a rich store of scrutinized possibilities, thus enhancing our capacity to choose wisely and well. I call this ideal of rigour for science, *aim-oriented empiricism*.

SCIENTIST: But surely you recognize –

PHILOSOPHER: One moment longer, please, and I will be done! Our next step involves asking the simple question: *why* do we seek *valuable truth*? Answer: to help people enhance the quality of their lives, realise their desirable human ends – help promote human progress, if you like. This means that in seeking to solve scientific, intellectual problems, we are in effect engaged in helping to solve personal, social problems, helping to realise desirable personal, social ends. Thus scientific, intellectual problems, in order to be understood properly, need to be seen as an aspect of personal, social problems. Scientific knowledge has a vital personal, social aspect, which needs to be clearly recognized if science is to be truly rigorous and objective. This even more rigorous ideal for science I call *humane aim-oriented empiricism*.

SCIENTIST: (For the moment giving up) Go on.

PHILOSOPHER: Back to *pure* aim-oriented empiricism (as we may call it). You remember the central idea: it is of enormous importance for science that we make the best possible choice of basic aims: this we are almost bound to fail to do, just because the aims are so profoundly problematic: here above all, then, we need to develop and explore possibilities, in order to enhance our capacity to choose wisely and well.

But all this is applicable whatever we may be doing! For, whatever we may be doing in our lives – whether on an individual basis, or a cooperative, social, institutional basis – it is all important that we make a good choice of basic aims (since if we make a *bad* choice of aims, choosing aims that either cannot be realised, or are inherently undesirable to realise, then we shall either have no success, or our success will be *undesirable*, tasting

66

like ashes in our mouths). In addition, however, for all kinds of reasons, it is more than likely that we will have *failed* to make the best possible choice of aims. Whatever we may be doing, at some point our basic aims are almost bound to become problematic. Therefore, if we are to give ourselves the best possible chance of achieving what we really want to achieve – our most realistic heart's desires – we need to articulate, explore, develop and scrutinize possible and actual aims for our activity or enterprise, so that we may provide ourselves with a rich store of scrutinized possibilities, thus enhancing our capacity to choose wisely and well. Unravelling of desire needs to be put at the heart of reason: for it is only if we have made a good choice of aim, an aim that really is desirable to realise, that reason can *help* us to attain what we want. If we have chosen *undesirable* aims, then the more rationally we pursue these undesirable aims, the worse off we shall be, the further we shall be from attaining what we really want to attain. Reason becomes a menace.

In brief: as a result of developing a more *rigorous* methodology for science (namely aim-oriented empiricism) we have arrived at a methodology which can be generalized to form a new desire oriented ideal for reason, of profound relevance for all that we do in our lives. Suddenly an entirely new and wholly desirable way of applying science to life becomes possible. Instead of merely applying the *products* of science to life, as we do at present, we can apply the *methods* of science to life, in a wholly humanly desirable way. Suddenly the possibility opens up of getting into our lives the kind of extraordinary, progressively successful character that is found within science, on the intellectual domain. Not only would science steadily progress: so too would art, education, politics, industry, international relations, and so on, in humanly desirable valuable ways. And perhaps even more important, desire oriented rationality can be used in our own personal lives, to develop our lives in desirable, fruitful directions, in directions which seem desirable and valuable to *us*. The heart of science, scientific method, reason, made more rigorous, becomes profoundly relevant to the most intimate aspects of our lives, to our hearts! That is the idea, anyway.

SCIENTIST: (With mock gravity) I see!

PHILOSOPHER: But all this is just a trivial preliminary to the really major reorganization of our ideas.

SCIENTIST: Oh!

PHILOSOPHER: With humane aim-oriented empiricism and aim-oriented rationalism before us, as just outlined, it becomes crystal clear that we can no longer conceive of science as something primarily pursued by *experts*, owned by experts, a product of the expert dissociated intellect or mind. Properly conceived, science is much too central and important a part of our lives to be thought of, and practiced, in such a way. In essence, science is *our* activity, *our* creation, the outcome of *our* concern. It is the outcome of our sharing of our concern for our world and for each other. It is a part of the expression of, and at the same time the outcome of, our concern to improve our relationships with the world and each other. The essential things, one might say, are *me, you* and *cosmos*: science is the adjusting of relationships between *you, me* and *cosmos*, so that these relationships become less painful, less frustrating, less restricting, more knowledgeable, more understanding, more appreciative, freer, more sensitive, more honest, more harmonious, more enjoyable, more trusting and loving. Obviously experts are important: some technical matters need to be delegated to experts, who may be permitted to pursue these matters under our kindly, watchful gaze, and with our help. But the *essential* thing is far too important, far too intimately associated with the very stuff of our lives, the very stuff of our personal identity, to be left to experts to decide upon. Science would not be helping us if *expert science* deprived us utterly of all free will, and was given a free hand to determine the very stuff, the very fabric, of our lives. There is no choice: we must say this: The centre of *gravity* of science (in a combined Bach-Newtonian sense) lies within our own hearts.

SCIENTIST: And who are you talking to now?

PHILOSOPHER: Anyone who might happen to be listening!

SCIENTIST: (With kindly irony, a sparkle in his eyes) All very impressive! But now I must dash. Perhaps we can meet in a day or two to go through your argument in more detail. I can tell you why I think it is a load of delightful, fanciful rubbish. But now, work calls.

PHILOSOPHER: Before you go, can I just sketch on the blackboard a kind of chart of the whole argument, so that it can be a kind of reference point for our future discussion?

SCIENTIST: I can give you two minutes. No more.

PHILOSOPHER: Quick, then. Blast! Where's the chalk?

(This is what he drew – diagram 1)

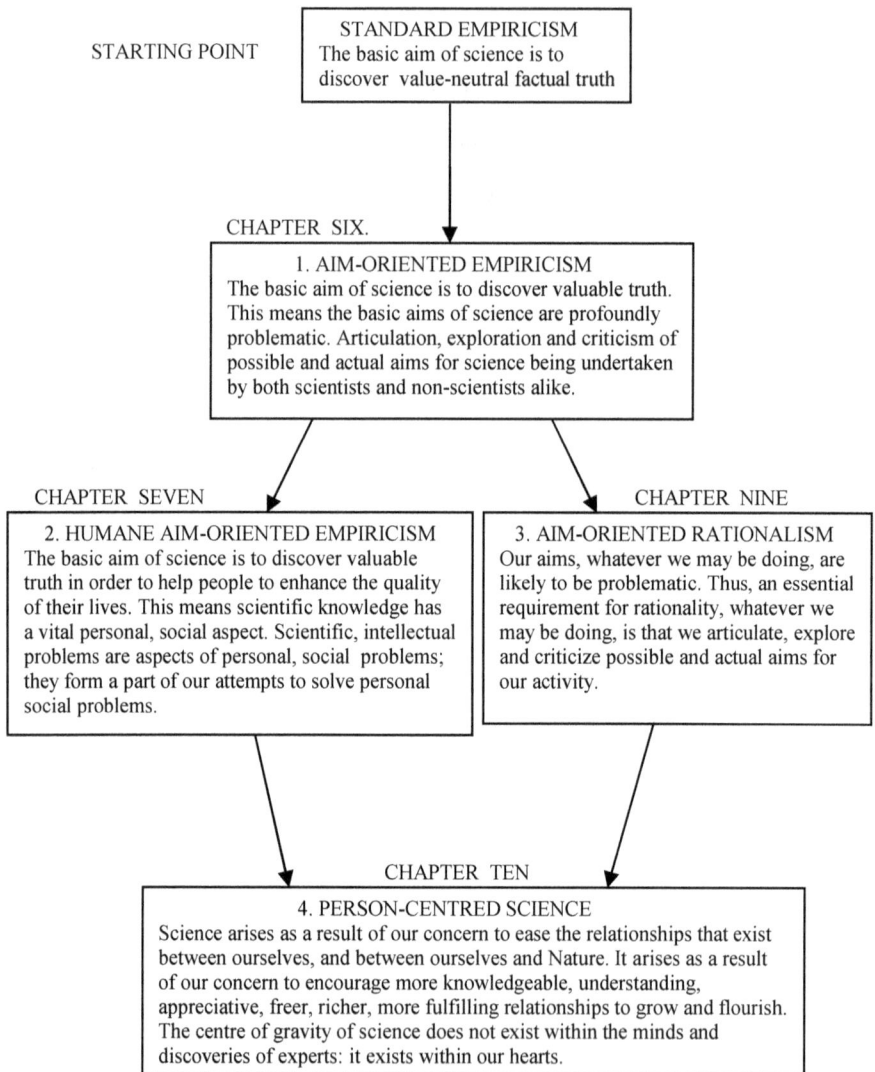

STARTING POINT

STANDARD EMPIRICISM
The basic aim of science is to
discover value-neutral factual truth

CHAPTER SIX.

1. AIM-ORIENTED EMPIRICISM
The basic aim of science is to discover valuable truth.
This means the basic aims of science are profoundly
problematic. Articulation, exploration and criticism of
possible and actual aims for science being undertaken
by both scientists and non-scientists alike.

CHAPTER SEVEN

2. HUMANE AIM-ORIENTED EMPIRICISM
The basic aim of science is to discover valuable
truth in order to help people to enhance the quality
of their lives. This means scientific knowledge has
a vital personal, social aspect. Scientific, intellectual
problems are aspects of personal, social problems;
they form a part of our attempts to solve personal
social problems.

CHAPTER NINE

3. AIM-ORIENTED RATIONALISM
Our aims, whatever we may be doing, are
likely to be problematic. Thus, an essential
requirement for rationality, whatever we
may be doing, is that we articulate, explore
and criticize possible and actual aims for
our activity.

CHAPTER TEN

4. PERSON-CENTRED SCIENCE
Science arises as a result of our concern to ease the relationships that exist
between ourselves, and between ourselves and Nature. It arises as a result
of our concern to encourage more knowledgeable, understanding,
appreciative, freer, richer, more fulfilling relationships to grow and flourish.
The centre of gravity of science does not exist within the minds and
discoveries of experts: it exists within our hearts.

DIAGRAM 1

CHAPTER SIX

AN AIM-ORIENTED IDEAL FOR SCIENCE

(They are at it again!)

SCIENTIST: I want to tell you, in plain terms, why I can't accept your argument.

PHILOSOPHER: Yes?

SCIENTIST: I see it like this. You have constructed a kind of soaring escalator of an argument. Put your foot on the bottom step, and effortlessly you are borne aloft to visionary heights – or to visionary fantasies! Everything depends on putting your foot on the bottom step, which I take to be to accept the idea that a *philosophy*, namely standard empiricism, exercises a profound influence over science. And it is just that idea which I cannot accept! Let me put it like this. *No philosophy whatsoever can be permitted to play any kind of really influential public role in science.* (Individual scientists may have their own *private* philosophies but that is quite another matter.) In fact, the thing can be put very simply like this: in science we do our utmost to ensure that nothing as intellectually disreputable as a *philosophy* – even a philosophy of science – can play any important rote whatsoever. The objectivity and integrity of science depend precisely on *excluding* philosophies from all influence within the intellectual domain of science. Evidence and logic alone operate: philosophies, along with almost everything else, are excluded (apart of course from being permitted a certain suggestive force for individuals, in the context of discovery). That is the whole point. In science we allow ourselves to be influenced only by the facts, the evidence, the data, experimental and observational results. All else is ruthlessly excluded.

You can see the trouble. Your soaring escalator does not exist, because the first step does not exist.

PHILOSOPHER: (triumphantly) Brilliant! Absolutely brilliant! With your usual intellectual perspicacity, you have put your finger on precisely the key issue.

SCIENTIST: (A little disconcerted at this reaction) What do you mean?

PHILOSOPHER: As you have just explained, your philosophy of science itself demands that philosophy – even philosophy of science – shall play no influential role in the public, objective, intellectual domain of science whatsoever. *The very thing which pervades all your thinking demands that no such thing as this must influence you in your thinking.* Once you accept your philosophy of science, it becomes impossible to criticize it, to reconsider it, within a scientific context. For if you do, you are no longer being scientific. You are no longer doing proper science: you are engaging in philosophy. Standard empiricism is a kind of intellectual lobster pot: once you are in, you can't get out; you even become unaware of the fact that you *are* caught in a trap. It's like Catholicism: entertain the idea that God exists, and you are caught: you can no longer doubt, because if God exists, it is a *sin* to doubt. If you do doubt, then this just shows how sinful *you* are, how unfit you are to decide these issues for yourself. Your only hope can be to put your trust in God.

You scientists behave in exactly the same way. The scientifically respectable thing to do – your philosophy of science sweetly assures you – is to forget about philosophy altogether, stop thinking altogether about *what it is you are doing*, and just get on mindlessly with the science. If you did start to think about what you were doing you would become, on the instant, scientifically guilty, a rubbishy old philosopher. Anything rather than that! And the fact that the poor fools who do try to think about what you are doing – the philosophers of science – simply can't make head or tail or it, and just come up with a lot of sterile rubbish, this, instead of *disturbing* you, just confirms you in your complacent decision that it is better not to think.

SCIENTIST: (angrily) I thought we had got past the stage of hurling abuse at each other.

PHILOSOPHER: Please! Don't be angry. It is not personal. All I am saying is that you think it is your scientific duty not to think seriously about the *philosophy* of your activity, about the aims and methods of your activity, and this very thought is what allows you to be victimized by a seriously *defective* philosophy.

Let me put the thing in this way. The whole source of the trouble can be put like this. Your very standards of scientific, intellectual integrity say to you: "It is wholly lacking in scientific, intellectual integrity (it is in fact something like the ultimate scientific sin) to look critically at scientific, intellectual standards of integrity". Consequently, you scientists, eager to be good scientists, don't look critically at these standards of scientific, intellectual integrity. Just that which utterly dominates all your thinking, is never called into question. It is as if scientists had decided that it was a scientific intellectual sin to call into question the idea that the earth is flat, or that the sun goes round the earth – although, of course, the thing is far, far worse than these trifles.

Really, all that is needed is for the scientific community to become aware of the extraordinary situation that has developed. In no time at all rush hour conditions will develop on what you have called my "soaring escalator of argument leading to visionary heights". There will be pandemonium. For once the first step has been taken, the thing is, in essence, so simple, so obvious, so elementary, such complete common sense, that anyone ought to be able to work it out for themselves in no time at all.

That is why I am so optimistic about the situation. All that is necessary is for our scientific fish to become aware of the fact that they are caught in the lobster pot of standard empiricism, and out they will swim, into the dark blue ocean. All that is necessary is for scientists to become aware of the fact that they are trapped in a straightjacket, and in no time at all the straps will loosen, fall away and our scientists will be able to stand on their feet, free men and women! At the moment you scientists are actively concerned to keep yourself trapped in your lobster pot, your straightjacket, because you have the absurd idea that it is your scientific intellectual duty to keep yourself there, and even worse, your scientific intellectual *duty* to keep at bay the mere *thought*, the mere *suggestion*, that this is where you are. The simple points that I am putting to you here must at all costs be kept out of the scientific journals, the textbooks, the university science courses. Good God! We must not start thinking and talking about anything as trivial, as uninteresting, as *intellectually disreputable*, as the fundamental aims and methods of science! Whatever next?

And it is just this which has prevented you from *improving* on your ideals of scientific intellectual integrity, your ideals of scientific rigour, as you have proceeded. You have not been able to do this because some time ago you reached the absurd conclusion: it is scientifically unrigorous to try to improve standards of scientific rigour.

SCIENTIST: I am trying not to get impatient. But you are not making it very easy.

PHILOSOPHER: I am sorry.

SCIENTIST: You see, I just am not aware of any "philosophy" dominating science. It just does seem to me to be the case that philosophy has no place in science.

PHILOSOPHER: Very well. Tell me, in very simple, elementary, non-technical terms, what you take to be the basic aims and methods of science, and why, in order to realise these aims, it is so important to keep "philosophy" out of science.

SCIENTIST: Well, it is all very boring and obvious. The fundamental aim of science is to improve our knowledge about the world, our knowledge about matters of *fact*. We do not have what you philosophers call, I believe, *a priori* knowledge. No one has a hot line to God. We do not *begin* with knowledge. The only way in which we can improve our knowledge about the world is by comparing our theories about the world itself via our *experience* of the world. We are obliged, in short, to submit our theories to completely impartial, objective, unbiased experimental testing. We choose those theories which best fit the facts.

None of this is to deny that scientists are *people*, who pursue science for all kinds of diverse *human* reasons – passionate curiosity, desire for fame, recognition, desire to make a successful academic career, even out of passionate concern for suffering humanity perhaps. All kinds of desires, ideals, pressures, may motivate the scientist. All kinds of factors – social, political, economic, ideological, cultural even – may *influence* individual scientists when it comes to choice of research topics, to choice of *aims* for research. Scientists may even be influenced by "philosophies", by "philosophies of science", in this kind of way. But in the long run none of this matters. For when it comes to considering whether a result, a theory, should be *accepted*, as embodying scientific *knowledge*, one consideration only is allowed to influence the thinking of scientists: does it do justice to the facts? Does it fit the evidence? Is it objectively more empirically successful than any of its rival theories? Or is the result or theory experimentally refuted? All other questions – questions concerning aims, personal motivation, method of discovery and so on – are completely and utterly ignored. And they *must* be ignored if science is to retain its objectivity, its integrity as science, its claim to produce authentic, objective factual knowledge. That is why no "philosophy" must be allowed to play any public, objective role in science. There – satisfied?

PHILOSOPHER: Completely: you have just given a very lucid exposition of the "philosophy of science" that I have been calling *standard empiricism*. And you have just made it crystal clear that this philosophy of standard empiricism completely *dominates* your scientific thinking.

SCIENTIST: (Suddenly very, very angry: his anger indeed surprises and shocks him with its intensity) So? What's wrong with that? What does it matter what it's called? (His tones are hard and sarcastic.)

PHILOSOPHER: (Quietly) Ordinarily it would not matter very much. It's just that in this case this "philosophy" which so completely dominates your scientific thinking is very, very seriously defective. And as I have been trying to point out to you, this is having serious repercussions for science itself, and indeed for *life*, just because of the immensely influential role that science plays in our world today —

SCIENTIST: *How* is it defective? *What* is wrong with it?

PHILOSOPHER: All kinds of things. A central defect is its complete dishonesty. It just is not the case for example that in science theories are selected solely on the basis of experimental success. This ignores the elementary point that an idea, in order to qualify as a "theory" at all, deserving any kind of scientific consideration whatsoever, has to have a certain simplicity, a certain conceptual coherence or unity. There is no impartiality in science. Messy, ugly, incoherent ideas are just never *considered*, whatever their empirical success might be if they were considered. The whole scientific enterprise simply presupposes, in short, that Nature, herself, is simple, unified, coherent – or at the very least, that she is such so as to behave as if she were like this, to a high degree of approximation. Metaphysical ideas about the world dominate scientific thinking in a way which standard empiricism does not even begin to acknowledge, explain or "justify". But please, let us leave this part of the argument for another occasion.

The most glaring and obvious defect of standard empiricism is, however, this simple, elementary point that I have been doing my utmost to get through your thick skull for some long time now, my friend. Standard empiricism quite grotesquely misrepresents the basic aims of science. Science does not seek to improve our knowledge of factual truth *as such*, as you assert. It seeks to improve our knowledge of humanly *valuable truth* – truth that is *beautiful*, fascinating, absorbing, that stirs our hearts, or truth that is *useful*, that helps us to solve our practical problems, realise our desirable human ends.

SCIENTIST: You are impossible.

PHILOSOPHER: Am I?

SCIENTIST: First, you give me a long rigmarole about the importance of simplicity in science – as if I was not aware of that! – and then tell me you don't want to discuss the issue. And then you accuse me of asserting something that denies that scientists are concerned to discover *valuable truth*. Do you think I need my head examining? Of course scientists want to discover "valuable truth" as you put it: nothing that I said denied that.

PHILOSOPHER: You do not listen to what I say. I did not accuse you of denying the importance of simplicity considerations in science. My concern was to indicate that you were quite unable to acknowledge the absolutely elementary *consequences* that flow from always giving priority to simple theories – from only even *considering* simple theories. Quite obviously it means that science must just presuppose that the world itself is simple (or at least behaves to a high degree of approximation *as if* it were simple). If you only considered *atomic* theories in science, then clearly this would commit science to the presupposition that the world is atomic in nature. Well, the same quite clearly goes for simplicity. All this stuff about choosing theories impartially in the light of evidence is just bunkum. Anything that clashes with your basic metaphysical convictions or presuppositions just never gets even *considered* in science, let alone put to the test, or anything like that. Your intellectual dishonesty is so unbelievably crude, elementary, obvious. A child could see it. And you, with all your erudition, your intellectual sophistication, all your knowledge and skill, your expertise, can't see it. Do you wonder when I tell you that you do science mindlessly, that you have simply stopped thinking about what you are doing? To put it bluntly: you are a fool. The sooner you recognise these elementary points about yourself, the better.

SCIENTIST: That's it. I've had enough.

PHILOSOPHER: (Jumping to the door, barring his way) For Christ's sake. Stay put. Let's work this thing out. I don't mind how much you yell at me, and nor should you mind how much I yell at you. But don't go. Don't run. Let's get this argument finished. I am not trying to *convert* you. I don't want to *persuade* you. I simply want you to *notice* one or two glaringly obvious points about your precious science which you seem at present utterly blind to. What you *do*, what you decide to *think*, is entirely *your* affair. But do please just open your eyes! O.K?

SCIENTIST: I can't see the need for hurling insults.

PHILOSOPHER: And I feel that you are completely oblivious to the insults that you hurl at me. It's all right for *you* to say, or to imply, that *I* am a fool to think there is something seriously wrong with science: you have all the authority of your church, Science, behind you. But if I start saying that you are a fool, and that a few things perhaps need to be sorted out in your precious church, then, oh, you become infinitely sensitive. You can tell others that *they* are stupid: but if anyone has the effrontery to tell you that you are stupid, precisely where it matters most to you, in your super intelligent activity of science, then, oh, that is quite another story.

SCIENTIST: Stick to the point.

PHILOSOPHER: I would like a plain answer to a plain question. Do you assess the intellectual progress of science in terms of the extent to which *valuable truth* is being discovered? Or do you assess the intellectual progress of science in terms of the extent to which truth *as such* is being discovered, whether valuable or trivial?

SCIENTIST: Scientists of course *want* to discover valuable truth. But assessing the intellectual progress of science has nothing to do with values.

PHILOSOPHER: I see. So a science which was amassing a vast store of irredeemably trivial, uninteresting factual truth would be making splendid progress?

SCIENTIST: Well, we might find such a science a bit dull. But, yes, strictly we should be obliged to say that from an intellectual standpoint the science was making progress.

PHILOSOPHER: And *values* would never determine what entered the body of scientific knowledge?

SCIENTIST: Of course not!

PHILOSOPHER: So if I began to count grains of gravel on paths, or leaves on trees, very accurately, and sent off my results to *Nature*, let us say, with a description of my methods of counting, and so on, I would meet with no trouble in getting my long series of papers published.

SCIENTIST: Don't be stupid!

PHILOSOPHER: (Mildly) Oh? I am being stupid am I?

78

SCIENTIST: (Doesn't reply)

PHILOSOPHER: (Still deceptively mild) Why am I being stupid?

SCIENTIST: Well, obviously in practice some kind of decision has to be made about what is worth publishing.

PHILOSOPHER: Oh, I see! Human values *in practice* influence what gets into the body of scientific knowledge. It's just *in theory* that values have no influence over the intellectual content of science.

SCIENTIST: Some kind of *minimal* judgement concerning significance obviously has to be made, or the pages of scientific journals would be cluttered up with all kinds of insignificant observations and results – as you rather tediously point out. But that in no way affects my main point. Scientists *want* to discover important truth – of course. Scientific, intellectual standards, however, have nothing to do with assessing the *value*, the *significance*, the *importance* of a contribution. That is an extra-scientific human judgement. The point is very simple Values are subjective. What is important or of value to one person may be trivial or positively harmful to another person. Science, however, must be equally acceptable to *everyone*. It must be *objective*. For this reason, we cannot have values creeping into the intellectual domain of science.

PHILOSOPHER: But if scientists *in fact* seek interesting truth, valuable, important, useful truth, then values are already there in science, in the choice of subject matter, in the choice of aims for science, even in the resulting content of scientific knowledge. Does not simple honesty – and also intellectual integrity, intellectual rigour – demand that this implicit evaluative aspect of science be clearly openly acknowledged, and rendered explicit, so that it can, for example, be discussed, criticized, reconsidered?

SCIENTIST: Rubbish! Scientists don't decide to accept results, theories, because of *values*. They accept them, as I have told you again and again, because they do the best justice to the facts, to the *evidence*. It is precisely when values, political or ideological considerations, enter the scientific domain that science is *corrupted*. Look at the Lysenko episode in Soviet biology. An absurd idea was upheld just because the idea seemed so *desirable*, so *valuable*. And the consequences were disastrous, not only for Soviet science, but also for agriculture, for food production.

PHILOSOPHER: (Dryly) Are you not making a rather elementary blunder here? There are two suggestions to consider: (1) Values are important in science because values influence scientists in choosing one theory from a number of rivals; (2) Values are important in science in that they influence scientists in their choice of subject matter, their choice of research aims, their choice of problems. The Lysenko episode is an example of (1). I have made it very clear, I would have thought, that I am putting forward suggestion (2). That's what gets me about you scientists. You shore up your idiotic ideas about science with such idiotic arguments. Your thinking about your subject really is screwed up. No, please, I'm sorry, I'm not just trying to be offensive.

Why don't you, just for a moment, shake out of your head these absurd tatty old *ideas* you have about science, and just look at science plain. Is it not precisely the great glory of science that it has discovered marvels about the world around us, that it has profoundly deepened our understanding, our appreciation of the world in so many ways? Surely science is important, of value, precisely because of the extraordinary capacity of science to make profound, beautiful, stirring, dramatic discoveries about the world about us, or discoveries that are of immense use, of great practical or utilitarian value. Think of all the great contributions to science, the dramatic advances in scientific knowledge, scientific progress. In every case, we judge these contributions to be substantial contributions to scientific knowledge, to the progress of science, not just because of the *quantity* of truth that they contain, but because of the significance, the importance, the value, the use of the truth we judge them to embody. Consider a theory such as Einstein's general theory of relativity. This is almost universally acknowledged to be one of the very greatest contributions to science. And yet the successful predictive power that it has over Newton's theory is not very great. Why is Einstein's contribution judged to be so substantial? Because of the profound *beauty* of the theory, because the theory brings together hitherto disparate elements – space, time, geometry, gravitation – in a wholly natural, unifying, extraordinary and novel way. We conjecture that the theory takes us an important step towards the ultimate goal of theoretical physics – to discover a unified pattern underlying all natural phenomena, to discover truth which we judge to be of value because it enhances our *understanding*, our appreciation of this extraordinary world in which we find ourselves.

The great glory of science is just that it does enable us to discover *valuable truth*. And because you have these absurd *philosophical delusions in your head*, you actually seem to be *ashamed* of just this glory of science. Nothing could illustrate just how blindly and dogmatically you cling to this shabby, tatty, philosophy of *standard empiricism*. Given the choice of acknowledging the true greatness of science, and of continuing to affirm your allegiance to your philosophy, it is your *philosophy* that you cling to. *Science* is disowned. You who claim to despise philosophy, nevertheless seem to cling to it like – like a frightened child to its mother's skirts.

SCIENTIST: I can assure you, these rather foolish emotional outbursts of yours sail right past my head. Let's hope they are therapeutic for *you* at least.

(There is a silence. Their eyes meet. Quite suddenly they both burst out laughing.)

SCIENTIST: Why are we being so stupid?

PHILOSOPHER: I don't know.

SCIENTIST: It's very odd. One can hardly think of anything more abstract, more unemotional, than questions concerning scientific rigour and rationality, and yet here we are, engaged in furious debate as if our very lives depended on the outcome.

PHILOSOPHER: Yes.

SCIENTIST: And it isn't even as if I care very much about the "philosophy" of science. It's *science* that concerns me, not the philosophy of science.

PHILOSOPHER: Me too.

SCIENTIST: Then why on earth are we getting so worked up, so angry with each other, about something that isn't really even our central concern?

PHILOSOPHER: A good question! I know why *I* tend to get angry. So often when I try to expound my ideas, I am rebuffed in such a casual fashion, the assumption being that I simply do not know what I am talking about, when in fact what I am saying has been completely misunderstood, and if anything it seems to me that the person meting out this dismissive treatment to me does not know what *he* is talking about. Foolishly, I get angry – which of course only makes the situation worse.

But another reason why we get so angry with one another, on occasions, is that the subject of our debate – scientific rigour – does after all *affect* what we both care about a great deal, namely science. A slight adjustment of public standards of scientific rigour ought to have *profound* consequences throughout the whole domain of scientific enquiry and indeed repercussions beyond science into life itself. If we really do care for science, then I think as a kind of offshoot we really ought to care for, to concern ourselves with, ideals of scientific rigour. That is really the basic point that I would like to get across to the scientific community. Ideals of scientific, intellectual integrity are really far too important, and play far too influential a role in science itself to be left to the amateurish, ignorant dabbling of *philosophers*.

SCIENTIST: Well, as a *general* point, that seems fair enough.

PHILOSOPHER: But there is, I think, another and perhaps much more fundamental reason why this issue of intellectual standards is so fraught with bitter emotional tensions.

SCIENTIST: Yes?

PHILOSOPHER: It comes from putting thought before life, instead of life before thought; science before reality, instead of reality before science.

SCIENTIST: What on earth do you mean?

PHILOSOPHER: Well, you won't get angry with me now if I just tell you, calmly, what I think about this? I promise: I am not trying to persuade you to believe what I believe.

SCIENTIST: (Laughs) Go ahead.

PHILOSOPHER: I suppose I see the thing in something like the following terms. Science as the outcome of people seeking, exploring, discovering the world, meeting fascinating, beautiful, valuable aspects of this mysterious world in which we find ourselves – science as the outcome of the open, warm-hearted sharing of this between people – that seems to me to be something moving and stirring. Dissociate science from a *personal* interest in aspects of reality, dissociate thought from personal enthusiasm, feeling, curiosity, and all kinds of bitter, unhappy, unacknowledged emotional drives take over. "To improve our knowledge of value- neutral factual truth" – what is there in that to stir the imagination, excite personal interest? Why should anyone want to pursue the aims of standard empiricist science? There is, of course, an answer! To make a contribution to science. To become *recognized* as a scientist. To win acknowledgement of one's *existence* on the intellectual map. To conquer death by making some immortal contribution to science, so that after one's personal death, one lives on, in one's contribution! One can at times almost conceive of science as a kind of gigantic Olympic games, with immortal fame as first prizes, Nobel prizes as rewards of a somewhat lesser rank, grading down through F.R.S.'s, scientific reputations, University chairs, Lectureships, Ph.D.'s, to the humble B.Sc.

The emotional centre of attention becomes: to win recognition; to achieve acknowledgement of one's existence as a mind, a reputable scientist. All scientists exist as persons, but only a few exist as "first rate minds", or whatever. One gets a hint of the fierce, bitter, unacknowledged *passion* that there is in all this from those well known, bitter priority disputes in the history of science. No wonder the disputes are bitter: one's very existence is at stake. What could be more natural, more desirable, than the desire simply to *exist*?

83

And, of course, the tragedy of all this is that almost everyone has to be bitterly disappointed. In the circumstances, it is no wonder that scientists, academics, insist that personal feelings and motivations have no rational relevance to thought, to science, to objective, intellectual issues. The personal emotions and motivations are so bitter, so hurt, so frustrated, so *angry*, so unhelpful, that the best thing to do in the circumstances is to keep emotion out of scientific, intellectual discussion. Keep the whole thing entirely impersonal and de-emotionalized. Given no real opportunity to find public expression, to be shared between people, the emotions, not surprisingly, become all the more bitter and frustrated – unless, of course, they die away altogether, leaving the cold, logical man of science, of mythology.

The idea that science, in order to be truly rational, ought to be openly passionate – well, in the circumstances, that idea is a non-starter.

You can see why, given all this, that standard empiricism should come to have such a powerful emotional appeal. (*Some* explanation is needed for the continued acceptance of standard empiricism, since standard empiricism certainly has no *intellectual* appeal.) Here is science, an opportunity for a very few to conquer death. Who is to decide who these few shall be? Who is to judge who finishes first? Or rather, who is to judge where the finishing line *is*? At all costs, we must have nothing uncertain or controversial about a matter of such decisive, central importance! Solution: let Nature herself decide for us! If she shouts: "No!" then throw the idea, the theory, on the scrap heap: the finishing line wasn't there. But if she murmurs "Yes"or "For the time being I am prepared to go along with the predictions of this theory", then crown the head of the genius concerned with glory. Nature, herself, has affirmed his (or her) greatness: it was not a human affair at all. The finishing line of the great race is set by Nature herself; she, herself, awards the prizes, as it were: the Nobel prize committee simply confirm, on a human level, what she has already pronounced.

And now along comes a mere philosopher – a person with no scientific standing whatsoever – with the suggestion that the rules of the game need to be changed, that *we* should begin to judge for ourselves what is of value, what is of importance, whether our concern is with scientific, intellectual issues, or issues of another kind, instead of handing the thing over to some authority – God, the Church, Nature, Science, Reason, the Nobel prize committee, the Scientific Establishment, Experts, Prophets, Gurus, or whatever. "What? What? Good heavens! How dare you come blundering into this area, of central concern to us scientists, but in no way something that can be a legitimate concern to *you*, a miserable philosopher, utterly ignorant of the practicalities of scientific research, with no scientific qualifications (and hence with no right to speak). I'm sorry: as far as we are concerned, *you do not exist*. Get back to your own miserable subject, philosophy, and leave us in peace to manage our own affairs. We are doing a damned sight better than you are, for a start. Put your own house in order, before you come troubling us with your fantasies!"

You can see the point. If *we* have to decide what is of genuine scientific intellectual value, instead of, as it were, pretending that Nature can decide for us, then everything becomes rather uncertain. We can no longer be sure that future generations will make the same kind of decisions that we make now. Our very opportunity of cheating death is imperilled. And even worse: perhaps we shall discover that we all make different decisions, different judgements! The overall unanimity of science may begin to crumble: science might begin to degenerate into a state somewhat like philosophy. Perish the thought. And how do we keep out the charlatans, the cranks, who believe that they have discovered the ultimate formula of the universe or whatever, and are tireless in their attempts to persuade everyone of the authenticity, the staggering significance, of their discovery?

SCIENTIST: Well, these are natural enough fears, are they not?

PHILOSOPHER: They are.

SCIENTIST: You are, of course, quite right in pointing to the phenomenon of scientific status seeking. But what can be done about it? Isn't the best thing just to forget about it, and get on with the really *interesting* pursuits?

PHILOSOPHER: Oh, you're so right! I knew we would be friends.

85

I would only add to that: surely we should also, as an offshoot of our central interest, concern ourselves just a little with the question of *why* we all seem to be, to a greater or lesser extent, caught up in this unpleasant, desperate, unhappy, bitter race for those miserably few prizes. Perhaps if we *think* about it, attend to it, with just a small pinch of honesty and realism, we might be able to work out a state of affairs that corresponds a little more satisfactorily to what we really all want, to what is really desirable. We might even be able to develop a science about which it would be very easy to have good, helpful feelings and motivations, which would help rather than hinder the intellectual aspect of science.

SCIENTIST: (Grinning) And what would you suggest? (*Sotto voce*) As if I didn't know!

PHILOSOPHER: One bit of the traditional scientific ethic seems to be: the *good* scientist is a modest fellow. Not for him worldly success. He is lost in absorbed interest in his scientific problems, with scarcely a glance over his shoulder at the social scene, the rat race and so on. When he is made a fuss of, he is surprised, bemused, bewildered. "Why me? I was only following up some matters I happen to be rather curious about. I was delighted to get a glimpse at what really interests me. But all this fuss, this bother, this fame? Quite frankly, I wish it would go away, and leave me in peace."

As a result of accepting that kind of picture of scientific moral rectitude, most scientists who are ambitious are probably led to feel somewhat guilty about their ambitiousness.

And yet, what could be more natural, more understandable, than the desire of a bright child, excited by science, to want to emulate Kepler, let us say, or Faraday or Einstein? We all know that these were wonderful men. Look at all the love that has been bestowed upon them – by posterity at least. Everyone wants to be loved – or, if one's ambitions don't go quite that high, at least one wants to be esteemed, admired, taken seriously, befriended.

And even the very desire to achieve immortality is surely only too understandable. It is the desire to *live*. It may indeed be the outcome of a quite basic biological drive to live (developed by natural selection) transformed by our unique awareness of death. Animals do not know they are going to die. But we do. Our whole life is spent in the desperate situation of a rabbit, cornered by a stoat, without escape, a deer about to be pulled down and slaughtered by wolves. We cannot ever really escape the knowledge that that is our situation, with a slightly extended time scale. This must undoubtedly create an entirely new situation for us, a new problem, unknown to animals, which must play havoc with our instinctive feelings, drives, impulses to act. Perhaps culture, art, myth, religion, knowledge, tradition, all arise out of our attempt to find some kind of solution to this appalling problem of death – the inevitable crumbling of all our acts, thoughts, feelings, hopes, dreams, products. If so, then the problem of death is a central emotional problem for science. It seems to me that a false – an unhappy – solution to this problem is, as it were, to flee from reality, from life, in horror, and escape into the safe, calm, immortal world of thought, of science. We will end up schizophrenics, one foot in science, the other in life, the two disconnected. And if we seek personal immortality in science, in thought, we shall not only feel guilty and be bitterly disappointed, in that in all likelihood we shall only receive the most meagre of footnotes in some abstruse historical work no one reads: even worse, we shall spend our life desiring our death for, of course, to want to be "immortal" in this sense, is to want to be dead.

A far braver, happier, more fulfilling solution to the problem is, I think, to adopt the attitude: there is nothing here to feel guilty about in this desire for immortality. For the problem is not that I am being too ambitious, but that I am not being ambitious enough. For my desire to be immortal springs from my more fundamental desire to *live*, and to find *love*. Let me put that first. By all means, let me concern myself with the future: but out of a sense of responsibility for our children, and our children's children. Are we not adults? Aside from that: no one decides my importance. *I* am the judge of that. I cannot leave it to others to determine what is of central significance to myself. Myself first, calmly, without guilt. And then consideration for others, as part and parcel of my concern to live. To desire to be a Kepler, a Newton, a Darwin, an Einstein, is altogether too paltry an ambition. I can successfully realise a far,

far more ambitious project than that! I can live. I can open my eyes, my heart, to this strange world in which we find ourselves. I can share my thoughts, my ideas, my discoveries, my problems, my experiences, my feelings, my suggestions, with others, in friendship. I can seize the central thing. Science, the whole of my world, can leap to life with my affirmation of life. I can find riches in life which will not be negated in their coming to be by their subsequent passing away. There are others, after all, who do things in this spirit. We exist *in any case*: we don't have to prove our existence. And with our intelligent affirming of our existence, so gradually more and more of what is of value in us can come into existence.

You see: I really do believe it. The centre of gravity of science is the human heart. At its most rational best, science is a passionate personal quest – shared between friends. The dead embers of science – the formulae, the experimental results, and so on – stir into life with the interest of a child. It is there for *us*, for *life*. We shouldn't let the *neurotic* hunt for life implicit in so much academic thought take that away from us. We need a new ethic for science – an ethic which puts life before thought, and which suggests that we should pursue thought in order to enhance life, not in order to try to escape from life.

SCIENTIST: (Smiling) What you have to say here, my friend, is I think not without a certain value. At least it deserves thinking on.

PHILOSOPHER: Oh! I am so glad to hear you say that. There are times when I feel it is *that* which comes close to being the heart of the matter.

SCIENTIST: I am amused, though. It seems that you really *do* think there is a kind of neurosis in science – a neurotic aspect that is. Neurotic, not in any technical methodological sense, but in a really rather straightforward emotional, motivational sense.

PHILOSOPHER: Well, my own personal view is – since that is what I am expressing for the moment – that the two things are intertwined with one another, mutually supportive of each other, the emotional and the methodological, that is. I believe that when this gets sorted out, as I am sure it will be before too long, then reason and emotion, thought and desire, instead of seeming to be at odds with each other, or operating in different universes, will harmoniously and magnificently cooperate. *Why* I think that will however, I suppose, only become clear, if at all, when we come to aim-oriented rationalism or *desire* oriented rationalism.

SCIENTIST: How do you think we should proceed now? I really do feel I don't want to wrangle with you any more. It seems pointless. Unnecessarily unpleasant. And actually destructive, in that we both become defensive and combative, instead of sharing with each other our real thoughts, concerns, problems. I really am sympathetically interested in some of the things you have been talking about. But that doesn't necessarily mean that I want to agree with you about everything.

PHILOSOPHER: Yes, I know; it's a problem. The only suggestion that I can think of just at the moment is that we proceed light-heartedly, as if it were all a joke, unrelated to big important issues, about which we naturally feel strongly.

SCIENTIST: How about this for a suggestion? We touch only very briefly on what seem, to the two of us, here and now, to be the central issues, the central problems. I indicate, very briefly, what seem to me the main difficulties with adopting your ideas; you indicate, very briefly, your reply; a gesture in the direction of a reply, and then we pass on to the next point. We don't have to assume that we have reached agreement at every step. But at least this will provide food for thought, so that we can perhaps both go off and develop our own ideas on all this. How does that sound?

PHILOSOPHER: The answer!

SCIENTIST: Good. Let's pick it up then where we left off when we got so idiotically angry with each other. As I understand it, your suggestion is at present this. Conventional methodologies for science in effect presuppose that the basic aim of science, from an intellectual standpoint, is to improve our knowledge of value neutral factual truth *as such*. The trouble with this "standard" approach, in your view, is that it completely misrepresents the real aims of science, the aims that science both does and ought to pursue. The real aim of science is to improve our knowledge of *valuable truth* – truth that is either of *cultural* value (in helping us to understand, appreciate the world, etc.) or of *utilitarian* value, in helping us to solve our human problems.

Your suggestion is: we need a methodology, a set of intellectual standards, criteria for progress in science, criteria for intellectual rigour, which are appropriate to the real aim of discovering valuable truth, instead of being appropriate to the phoney aim of discovering truth *as such*. Is that it?

PHILOSOPHER: That's exactly it. You really do have a brilliant capacity for picking things up I had no idea I had managed to communicate. The only thing I would add to what you have just said is this: As science advances, and as humanity advances – our social circumstances, needs, values and ideals evolving – so too ought the aims of science to advance, to evolve. Hence the methods of science ought to evolve as well. The unchanging methodology of science is on what might be termed the *meta-methodological* level, and consists of really very obvious, commonsense rules which stipulate how methods ought to evolve in sensitive sympathy with evolving aims.

SCIENTIST: I am not quite sure that I understand that. But I'll let it pass. My first problem: How can methodological rules, criteria of rigour, call them what you will, conceivably decide *for* us what is of value? Surely *we* must decide what is of value?

PHILOSOPHER: Of course! My suggestion simply amounts to this. Our methodology should help us to make the decisions that we really do want to make, instead of obscuring the nature of the decisions that lie before us, so that we are hampered in making the decisions that we really do want to make. Evaluative decisions, at present *implicit* in science, need to be rendered explicit, so that they can be discussed, criticized, reconsidered.

SCIENTIST: Second problem: Expert scientists may have a certain authoritative capacity to settle purely factual questions: but they can have no authority, no special expertise, when it comes to settling evaluative questions. How then can evaluative issues be a part of the intellectual domain of science, standards of scientific rigour having at least some relevance to helping us settle such issues?

PHILOSOPHER: My *deep* answer to you is that I think the idea of Science as an Authority, empowered to settle even purely factual questions for the rest of us, is obnoxious. Technical, detailed issues can perhaps be delegated to experts: but in the end *we* should decide; expert science should be there to help us reach good decisions, decisions we do really want to make.

My shallow immediate answer (independent of the above) is that of course scientists cannot decide evaluative issues for the rest of us. Precisely for this reason, it is absolutely essential, if science is to proceed in a truly rigorous, objective fashion, that the whole topic of possible and actual aims for science be thrown open for general discussion, within the community as a whole, scientists and non-scientists participating cooperatively in articulating, exploring, developing, criticizing possible and actual aims for scientific research.

If non-scientists do not participate in such discussion, and if scientists themselves decide what aims ought to be pursued, by means of the curious present mixture of individual decision, grant giving bodies, government directives settling broad issues of science policy, and private enterprise, then the great danger is that scientists will pursue aims that are of interest, of value to *scientists*, but not perhaps of so much value or use to the rest of humanity. Science will serve the subjective interests of scientists, instead of objectively serving the interests of people, of humanity.

SCIENTIST: But surely that danger is averted by the enormous diversity of research interests in science?

PHILOSOPHER: How do we know? We can only be sure of that if all kinds of possibilities are being actively and openly articulated and explored within the whole community. On the face of it, given the somewhat cloistered, sheltered lives lived by most academic scientists, given their somewhat specialized interests, the emphasis on the prestige of so-called "pure" research over applied research, the kind of emotional factors that tend to motivate research which we were talking about just now, one would be more than inclined to suspect that, at present, science does not pursue those aims of greatest value to people, to humanity. And a quite different kind of consideration leads me to adopt the same conclusion. In my view, a science which truly served the interests of humanity would take, as a first priority, the needs, the problems, of people facing the most adverse conditions. In general, today, these people live, in my view, in the third world. It is there that there exists the most appalling conditions of poverty, hunger, starvation, overcrowding, miserable living conditions, general hopelessness – or at least all the circumstances designed to produce hopelessness. In comparison, we in the technologically advanced countries live like kings, in conditions of luxury. It seems to me, however, that most scientific research today is pursued in technologically advanced countries, in order to further the interests of these countries. Technology, sensitively adapted to the needs and circumstances of the poor of the earth is not a first priority. And yet it is here, it seems to me, in this area of research, that our priorities should lie, that the Nobel prizes should cluster.

SCIENTIST: One last major difficulty. You say that science seeks *important* truth, not truth *as such*. In some large scale, overall sense, you may perhaps be right. But is there not a danger here in insisting that research scientists should only seek to discover important truth? Does not important truth often lurk behind apparent trivialities? Is not one major lesson to be learnt from the history of science that often some scientist pursuing some apparently trivial, abstruse problem, of interest only to himself, it seems, nevertheless by some miracle comes up with a discovery of profound significance, of immense importance? If we insist scientists pursue only aims that are clearly important, do we not run a very major risk of seriously sabotaging scientific progress? Does not the best hope for rapid scientific progress lie precisely in leaving scientists to pursue their diverse apparently trivial interests, in no way interfering with academic freedom, just so that we have the best overall chance of hitting on that which is of real importance?

PHILOSOPHER: First, what I have to suggest in no way interferes with academic freedom. On the contrary, I believe it enhances it.

Second, your point concerning the problematic character of the aim of discovering important truth is the central insight behind my suggestion. My suggestion is intended to help scientists to hit upon these ostensibly trivial little problems behind which major discoveries lurk. It is, in short, amongst other things, a rational (though of course fallible) method of *discovery* in science.

SCIENTIST: Well, well! Let's have it, then.

PHILOSOPHER: It can be developed like this. First point: Would you agree that *rigour*, whatever else it may be, involves at least making explicit, and so criticisable, that which is implicit, influential and problematic?

SCIENTIST: How do you mean?

PHILOSOPHER: Well. Suppose there is an argument, a proof if you like, and you discover that in the argument there lurks an implicit assumption which is *influential*, in that if you reject it, the conclusion of the argument is drastically affected, and also *problematic* in that it is not at all certain that the assumption is true. In making explicit this implicit, influential and problematic assumption, laying it bare for inspection, adding it to the premises of the orpiment for example, are you not considerably enhancing the rigour of the argument, the proof?

SCIENTIST: Of course!

PHILOSOPHER: Now my point is this. Once we recognise clearly that the fundamental aim of science is to discover valuable truth, two things instantly leap to the eye: assumptions that scientists make about what more or less specific aims to pursue will exercise a profound influence over science; such assumptions must inevitably be profoundly problematic. Rigour demands then that these influential, problematic assumptions concerning aims be made thoroughly *explicit*, precisely so that they can be scrutinized, developed, modified. At present, by and large, this does not go on within the intellectual domain of science. Why not? Because standard empiricism demands – in the interests of "scientific rigour" of all things – that it shall not go on. According to standard empiricism, the basic aim of science is simply to discover truth *as such*. Rigour demands that only experimentally testable theories, and experimental and observational results, be allowed to enter into the intellectual domain of science. All the rest must be excluded. Thus, as a result of the attempt to make science conform to the ideal of rigour of standard empiricism, assumptions about aims, instead of being made explicit, are forcibly kept in an implicit, concealed, hidden state. The attempt to make science conform to standard empiricism actually sabotages scientific rigour. Now do you understand all my wild talk about neurosis, lobster pots and straightjackets?

SCIENTIST: Remember. You are not trying to *convince* me. Only to interest me.

PHILOSOPHER: I'll try to remember it. But please do let me develop in just a little more detail the argument just given. It is, I believe, *so* important for developing a truly objective, rigorous science, sensitively responsive to human needs, problems, desires, aspirations.

SCIENTIST: I will do my best to check my impatience.

PHILOSOPHER: First point. Our choice of aims for science exercises a profound *influence* over science. It influences the rate of progress of science, since if we choose aims that can be scientifically realised we shall make rapid progress, whereas if we choose aims that are unrealizable we shall make no progress. It influences the subsequent content of scientific knowledge, in that it influences what we choose to develop knowledge about. And it influences how responsive science is to the needs, the problems, the aspirations of *people*.

SCIENTIST: Yes.

PHILOSOPHER: Second point. It must remain permanently and profoundly problematic to choose the best possible aims for science. In what directions do really valuable, scientifically discoverable truths lie? We do not really know: this concerns the domain of our ignorance. What is it important to try to discover, in any case? Important for whom? What will be important in thirty, one hundred, five hundred years' time? Clearly, the chances of our discovering the very best possible aims for science are almost infinitely remote.

SCIENTIST: Very well.

PHILOSOPHER: *It is of fundamental importance for science, and for humanity, that we make the best possible choices of aims for science; and it is almost inevitable that we will fail to make the best possible choice of aims for science. Here, above all, then, surely, we need to proceed intelligently, critically, imaginatively, wisely.* We need, above all, accurately to *articulate* the aims that we are at present pursuing; and we need to develop alternative possibilities; we need imaginatively and critically to explore and scrutinize aims that might be pursued. In this way we can provide ourselves with a rich store of critically examined possibilities, thus, hopefully, enhancing our capacity to choose wisely and well.

Let me quickly draw two diagrams to pin-point the differences between these two ideals for scientific rigour, *standard empiricism* and *aim-oriented empiricism*. (He draws diagrams 2 and 3.)

You can see the point. The best aims for science lie in the highly problematic region of overlap between the scientifically discoverable and the humanly desirable. In order to give ourselves the best opportunity of discovering this problematic region of overlap we need to articulate both our guesses as to what is scientifically discoverable and our guesses as to what is humanly desirable, humanly valuable. We need both considerable scientific knowledge and understanding and a sensitive and intelligent insight into the needs, the problems, the feelings, the desires and aspirations and ideals of people. These two kinds of knowledge and understanding need to be brought into intimate communication with one another.

It may be that only members of the professional scientific community can be expected to possess the specialized scientific knowledge that's needed in order to make a good guess at the scientifically discoverable. But I am, myself, inclined to doubt this. We need to remember that in attempting to discern scientifically realizable aims for science we are attempting to see into the domain of our ignorance, into the region of the great unknown. It is just here that we do not have elaborate, specialized technical knowledge. The ideas, proposals, conjectures, guesses,

Influenced by human needs, emotions and desires, political objectives, moral consid-erations, ambition, curiosity, economics, commercial interests, national prestige, international conflict.

→ Extra scientific, human, social aims of science ←

Ideas for future scientific theories, hoped for technological discoveries and developments, hoped for explanations for inexplicable phenomena, hoped for solutions to existing scientific problems.

Rift between the 'scientific' and 'extra-scientific' aspects of science

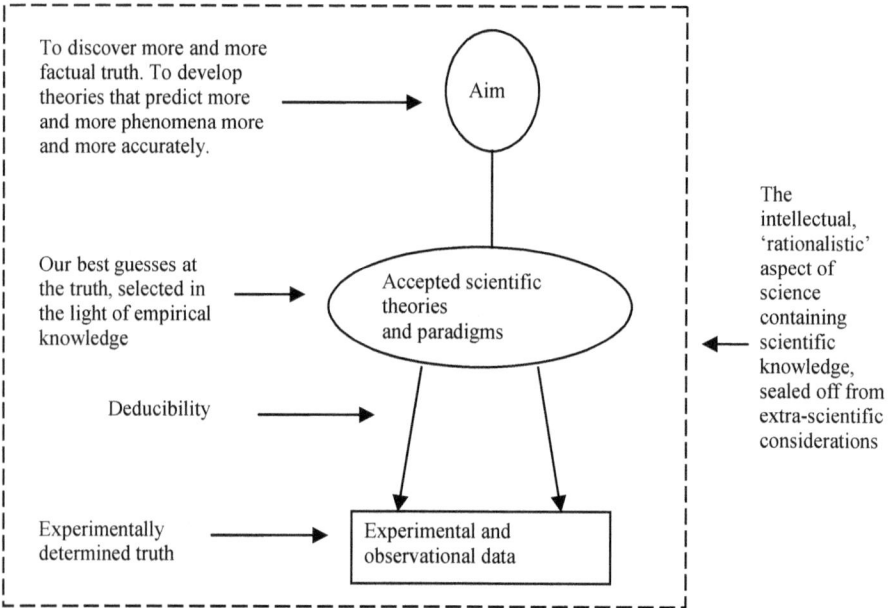

To discover more and more factual truth. To develop theories that predict more and more phenomena more and more accurately. → Aim

Our best guesses at the truth, selected in the light of empirical knowledge → Accepted scientific theories and paradigms

Deducibility →

Experimentally determined truth → Experimental and observational data

The intellectual, 'rationalistic' aspect of science containing scientific knowledge, sealed off from extra-scientific considerations

Standard Empiricism

DIAGRAM 2

96

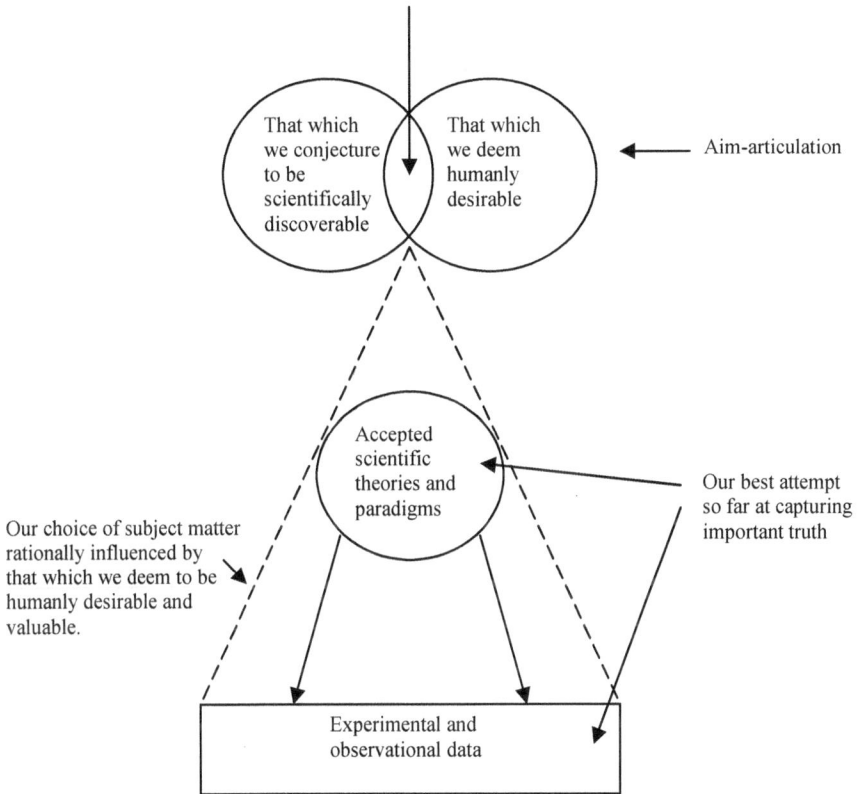

That which we conjecture to be the best aims for science: important, discoverable truth.

That which we conjecture to be scientifically discoverable

That which we deem humanly desirable

Aim-articulation

Accepted scientific theories and paradigms

Our choice of subject matter rationally influenced by that which we deem to be humanly desirable and valuable.

Our best attempt so far at capturing important truth

Experimental and observational data

Aim-Oriented Empiricism

DIAGRAM 3

97

speculations, bubbles, that we need to discuss in an attempt to discern the scientifically discoverable will often of necessity be somewhat vague and informal, and thus *necessarily* understandable to the non-expert. It is just here, in other words, in this crucial domain of possible scientifically realizable aims for science that it could well be feasible for scientists and non-scientists to communicate with mutual comprehension. The failure to articulate our best guesses about the domain of our ignorance hides from view the one part of scientific knowledge that ought to be generally comprehensible without specialized knowledge. It ought to be possible for non-experts to participate usefully and valuably in this vital area of the "foundations of science" as it is usually called (although, in my view, it would be better to call it the "apex" of science). Failure to articulate clearly our guesses about the domain of our ignorance serves quite unfairly and unnecessarily to keep non-experts excluded from, and in ignorance of, the true nature and excitement of the scientific quest.

Be that as it may, it is quite clear, as you have yourself pointed out, that the scientific community cannot claim to have any special gifts, skills, authority or expertise when it comes to discovering the other aspect of the best aims for science, namely the humanly valuable. We can only hope to articulate the humanly valuable in an objective, unbiased way as a result of open, public debate and discussion within the whole of society.

In brief, if we are to give ourselves the best chances of developing a science that pursues good and wise aims, aims that are both scientifically realizable and humanly valuable, then it is essential that the whole topic of aims be thrown open for general public debate and discussion within the community as a whole, so that there can be a genuine flow of ideas, suggestions and criticisms, genuine understanding, in *both* directions, between the scientific community, and the broader community within which science exists. Scientists cannot determine for the rest of us what the aims of science ought to be: for that would almost inevitably lead to a science pursuing aims of special interest to scientists – a *subjective* science oriented around the subjective interests of the scientific community. The rest of us cannot determine for the scientific community what the aims of science ought to be for that would almost inevitably lead to a science pursuing scientifically illiterate aims. It is only if there is mutually cooperative communication, discussion and criticism between scientists and non-scientists, both sides being willing to learn from the other, that there can be any hope of a truly objective science – a science sensitively and intelligently responsive to both scientific knowledge and human needs and aspirations.

SCIENTIST: But would not this kind of thing lead to the neglect of *pure* research, which has no immediate human value?

PHILOSOPHER: My friend, I am ashamed of you! How could you say such a thing after all that I have been saying? Have I not constantly been at pains to emphasize that science is of human value in two rather different ways? First, science is of value in that it can be used to increase our – *people's* – knowledge, understanding, appreciation, of the world around us; second, it is of value in that it can be used to help us to solve our practical problems, realise our personal, social objectives. (In the end, of course, these are but two slightly different aspects of the central thing: improving our relationships with the cosmos and with each other.)

Thus, there is no such thing as pure research. All research is applied – either culturally or technologically. All research, in one way or another, properly understood, is a part of the broader human endeavour to discover and create meaning in our lives, alleviate suffering, unfold the enjoyment of life.

It is, however, important that we try to distinguish research that has *cultural* value from research that has *technological* value. It is, of course, very satisfactory if a piece of science is simultaneously of value in both kinds of ways. But this may not happen. Einstein's general theory of relativity is of profound cultural value, but seems to be of no conceivable technological value; new drugs are often of immeasurable technological, utilitarian value, and yet are of little cultural value, in that they may not represent, or arise from, an increase in understanding of much note.

One of the more pernicious effects of the widespread acceptance of standard empiricism arises precisely from the fact that progress in science is defined in a sense which fails to distinguish between cultural and technological progress. As a result, the progress that is achieved often falls between two stools, neither one thing nor the other. This is most marked, I believe, in contemporary physics. Physicists, with the customary unthinking idiocy of the scientific community, assume that progress in physics means: developing theories which progressively predict more and more phenomena more and more accurately. As a result, they lose interest in the real *cultural* task of theoretical physics – natural philosophy – namely: to enhance our understanding, our appreciation, of the universe. In order to enhance our understanding, our appreciation, of the universe we need, of course, to test our ideas, our theories; but simply to be able to predict more and more phenomena more and more accurately does not in itself equal understanding. Our theories must attempt to delineate aspects of the nature of reality; they must not just be algorithms, which predict phenomena successfully, but remain silent on the question of what is really there. Again, as a result of upholding the standard empiricist conception of scientific progress, physicists tend to lose sight of the *technological* tasks of physics: the phenomena that are predicted tend not to be especially relevant to the needs, the problems, the frustrations, the difficulties of people. On both counts, physics degenerates into a kind of sterile, technical, meaningless game, absorbing vast sums of money, manufacturing some scientific reputations and Nobel prizes no doubt, but otherwise without human meaning or value. It fills me with intellectual and moral horror. The thing is almost like a very expensive form of academic philosophy – with the same sterility, the same essential mindlessness. And now let me at once add: *of course* I exaggerate. It is not as bad as I make out. Nevertheless, I do believe physicists have abandoned the heart of their subject. You can see it in the following simple fact. If one is really concerned to understand Nature – as Einstein was – then an absolutely fundamental problem, of central significance for such an undertaking, which cannot possibly be avoided, is simply: Well, what sort of things are these particles – electrons, protons, etc. –

which appear to have both wave-like and particle-like properties? No one with any real curiosity would want to avoid finding an answer to that central mystery. What does one find? The physical theory that deals with this matter, quantum mechanics, is immensely successful empirically even though it is completely silent on the crucial issue: are electrons waves, particles, or what? Thus quantum mechanics fulfils all the conditions for an acceptable theory demanded by standard empiricism. At a stroke it has become unscientific, metaphysical, philosophical, mystical, to ask the elementary and obvious question: But what sort of things are these wretched electrons? Generations of physicists are trained, brainwashed, into thinking that the problem is meaningless (the brainwashing going by the name of education). So dogmatically gripped are they by a *philosophy* of science that they abandon the heart of their subject. They proceed dutifully and mindlessly with the set task of predicting more and more phenomena. "Has the subject been taken over by robots, or what?" one almost feels like asking.

Let me hasten to add that I am not alone in thinking all this. This is what Einstein wrote in a letter to Schrödinger. (Our Philosopher snatches up a little book to quote from.)

"You are the only contemporary physicist, besides Laue, who sees that one cannot get around the assumption of reality – if only one is honest. Most of them simply do not see what sort of risky game they are playing with reality – reality as something independent of what is experimentally established. They somehow believe that the quantum theory provides a description of reality, and even a *complete* description; this interpretation is, however, refuted, most elegantly by [you] ….. If one wants to consider the quantum theory as final (in principle), then one must believe that a more complete description would be useless because there would be no laws for it. If that were so then physics could only claim the interest of shopkeepers and engineers; the whole thing would be a wretched bungle." [1]

Our contemporary physicists, however, are happily content with this "wretched bungle".

Twenty-two years earlier, Einstein had written, again to Schrödinger: "The Heisenberg-Bohr tranquilizing philosophy – or religion? – is so delicately contrived that, for the time being, it provides a gentle pillow for the true believer from which he cannot very easily be aroused. So let him lie there." [2]

They are still sleeping. But, frighteningly, it is not just the Heisenberg-Bohr philosophy that has put the physicists to sleep. It is the tranquilizing philosophy of standard empiricism that has put to sleep the science community *en masse*.

As just one indication that the physicists are still asleep nearly fifty years later, let me give just one further quotation: In 1970, amazingly, a physicist, Ballentine, actually had the nerve, the audacity, to raise again the central problem: Well, what does quantum mechanics really tell us about these little electrons and protons and so on? This is what the editor had to say:

"The subject of the following paper lies on the border area between physics, semantics and other humanities" [3]

You can see the point. The central and fundamental problem of natural philosophy belongs to the humanities, to semantics. (Clearly the editor could not bring himself even to utter the word "philosophy".) Given the choice of pursuing the central and wonderful aim of natural philosophy – to improve our understanding of Nature – or of sticking dutifully to the edicts of the philosophy of standard empiricism, the physics community does not hesitate: with a few noble isolated exceptions, the physics community disowns science, disowns reality, and clings to its *philosophy*!

SCIENTIST: But perhaps physicists are tackling this fundamental wave/particle duality problem. It's just that they don't know how to *solve* it.

PHILOSOPHER: I wish I could agree with you, but I cannot, for the following simple reason. I am, myself, no physicist. My interest in physics is only a spin-off of my interest in reality. And yet, lacking all technical knowledge and skill, unable even to write down the Schrödinger equation (a central component of elementary quantum theory), let alone solve it, I have myself concocted an

Idea for a solution to the problem. Could I get the Idea published? Of course not! The Idea was not physics, in that it predicted no new phenomena. It was not philosophy of science, in that it did not analyse existing physical theories. No, it was only an Idea as to what sort of things electrons might really be, an Idea for a new kind of micro realistic quantum theory which would not be "a wretched bungle". And of course who would be interested in a stupid thing like that? After about six years of vain efforts to try to get the Idea published, it suddenly dawned on me how I could make out a case, at least, for saying that the Idea is experimentally testable. The thing had become conventional physics: it has now been accepted for publication.[4]

You can see my point. I, a mere bystander, an ignorant amateur, have come up with an idea for a solution which no one in the physics community has bothered to think up, or develop, during fifty long years. I do not think that this indicates a very lively interest in the problem.

Incidentally, you know the widely held view that contemporary physics has become irredeemably incomprehensible to the layman? The reason is, I believe, quite simple: physicists themselves no longer really understand what they're doing. They have abandoned the attempt to understand. Even worse: they hold it to be their duty, as good standard empiricist scientists, to abandon all *attempts* to understand. "One understands quantum theory when one understands there is nothing to understand", as one pundit on the subject wrote fairly recently. Instead of there being comprehensible Ideas, understandable informally, and needing only to be dressed up in formal mathematical language for purposes of experimental testing and appraisal – instead of this, there are *only* mathematical formulae, devices for predicting phenomena. To understand is to know how to work the mathematical machinery. Of course, the lay person cannot understand.

In short, modern theoretical physics is incomprehensible because it is made to conform, as far as possible, to the incomprehensible philosophy of standard empiricism.

SCIENTIST: You are breaking our agreement. You are going angrily on and on, and I am losing the drift.

PHILOSOPHER: Yes, I am sorry. I only wanted to give just one example of how the attempt to make science conform to the philosophy of standard empiricism has perverted the content of scientific knowledge, and made it seem that personal interest in science from a "cultural" standpoint must be impossible for everyone except the experts. If anything, it's all the other way round. Only the amateurs, perhaps, can keep alive love of natural philosophy –

SCIENTIST: Please!

PHILOSOPHER: I am sorry.

SCIENTIST: Could we return to this proposal of yours of scientists and non-scientists sharing cooperatively in articulating actual and possible aims for science?

PHILOSOPHER: Yes.

SCIENTIST: How do you imagine it corning about? In practical terms?

PHILOSOPHER: One possibility would be to have a number of popular journals, available in news-stands, concerned solely with articulating, exploring, developing, criticizing possible aims for various sciences. One for science as a whole. One each for the natural sciences, the biological sciences, and the social sciences. And then, perhaps, a few more specialized journals for individual scientific disciplines.

The idea would be to encourage a kind of riot of delighted, free, wild, imaginative exploration of possibilities. Any suggestion, however wild and speculative, as to what important truth *might* exist to be discovered would be acceptable. The only restrictions on publication would be that, in order to be acceptable for publication, a paper would have to make a genuinely novel suggestion (or develop in a novel way a previous suggestion); and it would have to be entertaining to read. Criticisms of previous suggestions would also be acceptable. Accurate, non-technical accounts of aims *actually* being pursued by science would also be published. If other restrictions begin to seem desirable, as aim articulation proceeds, then they can, of course, be introduced as the need arises.

SCIENTIST: All very delightful. It seems, however, more like a relaxation of rigour, rather than a tightening up of rigour.

PHILOSOPHER: (In a good humour) Oh! You haven't been listening to my argument. Rigour isn't *rigor mortis* you know. What I am indicating enhances the rigour of science because, to repeat, it renders explicit, and so criticisable, that which is at present largely implicit, influential and controversial. You can see the fundamental importance of distinguishing clearly the three categories of contribution to the intellectual domain of science: (1) experimental and observational results, (2) testable theories, (3) aims. Quite different criteria operate within these three distinct categories. We need to conceive of our scientific knowledge as being composed of our best ideas at all three levels. Our knowledge includes not only what we know, but also our best guesses about what we do not know, but hope and desire to discover.

SCIENTIST: One thing worries me about your suggestion. It seems to me, the world being what it is, that all kinds of cranks would get *their* wholly useless ideas published, whereas scientists with good ideas would keep them to themselves, until they had had an opportunity to develop them into an acceptable *scientific* form.

PHILOSOPHER: It is precisely for this reason that aim articulation must be taken seriously as an integral part of the intellectual domain of science. An idea proposed in an aim articulating journal ought to win its author a joint Nobel prize, let us say, if it is subsequently developed into a piece of valuable scientific knowledge.

SCIENTIST: But that would mean that any old fool – –

PHILOSOPHER: Exactly. You have completely grasped the idea. It should be possible for any old fool to win a Nobel prize. Just as long as the idea is a good one. It is absolutely essential that aim articulation, though open to the general public, is nevertheless taken absolutely seriously by the scientific community, as a vital,

integral part of the intellectual domain of science. A scientist needs to have some knowledge on all three levels: experiment, theory and aims. Scientific textbooks and monographs need to include discussion at all three levels. School and university courses need to include classes and lectures on all three levels. Examinations need
– –

SCIENTIST: Yes, yes, I get the general idea.

One thing still bothers me. Why should we have any reason to suppose that this process of aim articulation would ever generate any useful ideas? The domain of our ignorance is, after all, just that: the domain of our ignorance. Aim articulation could only amount to pure, random guesswork.

PHILOSOPHER: But that ignores that we are never *completely* ignorant of the domain of our ignorance. Or rather: our common sense assumption that we do possess *some* knowledge commits us inevitably to the assumption that we have some kind of vague knowledge about all that there is.

SCIENTIST: What?

PHILOSOPHER: Think of it this way. Suppose we divide up reality, all that there is, into two rough categories; that which we know (more or less), and that which we do not know, that which we are ignorant of.

SCIENTIST: Yes.

PHILOSOPHER: Well, it is surely fairly obvious that it is only if we assume that we have some kind of vague knowledge of the domain of our ignorance that we have any right to assume that we possess any genuine scientific knowledge at all. For unless we assume that the domain of our ignorance is reasonably stable and well behaved, with respect to our ideas of what constitutes stability and good behaviour implicit in our present scientific theories, we can have no reason to assume that this domain of which we are ignorant will not suddenly intrude upon the domain of our knowledge in some violent, unruly fashion, entirely disrupting the straightforward predictions of our present-day scientific theories. Even our most trivial, commonplace assumptions of knowledge involve implicitly assumptions of some kind of vague knowledge of all that we are ignorant of, the whole cosmos, the ultimate nature of reality. My assumption that I *know* I can walk across a room implicitly presupposes that I know that there is not, for example, some vast cosmic convulsion beginning even now at the other end of the universe, which will spread with near infinite speed to engulf the earth, this room, me, before I have had an opportunity to take three steps. Even my assumption that I know that I can wave a hand, smile, speak, think – my assumption that I know I *exist* – presupposes some kind of vague knowledge about the nature of ultimate reality. For in assuming that I know these things I in effect assume that I know that the ultimate nature of reality, whatever precisely it may be like, is at least such that it is possible for people, on occasions, to do things, act freely, have free will. The fact of the matter is that we live, breathe, act and have our being within the cosmos, and as an aspect of ultimate reality, even if we may pretend to ourselves that we live and breathe within a somewhat more restricted environment – within this room, on this street, in this field, amongst things that I can see and touch and know, by means of my ordinary, restricted, personal experience.

SCIENTIST: So according to you we are citizens of the cosmos; we have cosmic stature, as it were?

PHILOSOPHER: I am serious! We really do have cosmic stature. We can either ignore it, pretend that we are not citizens of the cosmos, like the animals. Or we can recognize it, acknowledge it: "Hello Cosmos. Wow! You are big. And I am so absolutely *minute*, almost invisibly small. Well never mind. I refuse to be intimidated. What's it all about then, Cosmos? What's going on here? Speak up, I can't hear you. Oh I see, you're the sort of thing that doesn't speak, that doesn't have ideas about what it's all about. What does that mean? It must be up to me to decide! How extraordinary. This little minute speck of *me*, this tiny, perilous fragment, to decide what it all means, for old unknowing Cosmos itself even. What an extraordinary thing. How delightful. What a responsibility! Now, let me see, what shall I decide it's all about? Oh dear, that's a problem. I wonder what conclusion other people have come to? Good heavens! What a confused babble of voices. Everyone so convinced that they alone have the answer, everyone else being wrong. They all seem to be so impatient with one another, so angry with one another. They're not really listening to each other. They certainly don't seem to be delighted with all these quite peculiar answers. But they can't all be right – at least not in all details. Surely one can at least say this? Something is there, beautiful, lovely: *life!* Or rather, there are lots of things that are beautiful, lovely. I can feel it within me, this quick urgent desire to meet, to experience, loveliness. I'll nose it out. That's what I'll do. Since old Cosmos is permanently asleep, a great old snoring dog, I'll have to find out what the regular rhythm of his breathing allows that is of the loveliest, and seek *that* out. And of course I'll probably get into all sorts of stupid tangles, make a mess of things, get the idea that what is really desirable is for ever locked away from me and so on. Never mind. Expect all that. Don't get too over-excited, too disappointed, too desperate, and I'll probably be all right. Stupid old Cosmos: you are very big, and you have been around a long time. *But I am alive!* I wonder, do other people have these crazy one-way dialogues with the Cosmos?"

There! *That's* the sort of attitude we should have, I think. At any rate, that's the sort of attitude *I* have. It seems to me a child of three or four ought to be able to have a conversation with the cosmos like that. There should be adult encouragement, the adult support and openness to make such a thing possible. And look what a mess humanity, by and large, has made of such a conversation! Some steps have taken thousands and thousands of years to work out, slowly and painfully. Other steps have not been worked out even yet. Such simple things buried for ever beneath a morass of pomposity and lies.

SCIENTIST: I'm sorry, all this may be very true, but what relevance does it have to science?

PHILOSOPHER: But my whole point is just this: science, at its rational best, is the outcome, the expression and sharing, of just such a dialogue with the cosmos. If the dialogue is to proceed in truly rigorous fashion, then we need to render explicit our vague conjectural knowledge about the whole of reality that is inevitably implicit in the scientific knowledge that we recognize today as "knowledge". We need to exploit *all* our knowledge. Consider theoretical physics. The whole enterprise of theoretical physics presupposes that some degree of order, lawfulness, harmony, unity exists in the universe, not utterly different from our present-day ideas of order and lawfulness – explanations for ostensibly diverse, complex, disorderly phenomena thus being possible to discover. Theories in physics are only acceptable, indeed only qualify as "theories" at all, insofar as they comply, to some extent, with the presupposition, the conjecture, that order exists in the universe. For in order to be acceptable in physics a theory must not only meet with empirical success: it must also possess some degree of internal unity, harmony, coherence; it must be such that it attributes a harmonious pattern to phenomena. Metaphysical presuppositions about the nature of that which we are largely ignorant thus play an essential, if at present somewhat implicit, role in the selection of theories in physics. Our very choice of terminology in theoretical physics involves implicit metaphysical presuppositions. If we are to proceed rigorously and rationally, we need to make explicit these implicit, influential problematic presuppositions, precisely so that they can be criticised, developed, and hopefully, improved.

As our knowledge improves, so too our tentative knowledge of that of which we are as yet largely ignorant, but hope and desire to discover, also improves. Our *aims* improve: and as a result our *methods* improve. In theoretical physics as it exists today one sees very strikingly exemplified the *beginnings* of this sophisticated process of aims and methods being developed in harmony. Of central importance in theoretical physics today is the development of invariance and symmetry principles, which may be interpreted as tentative methodological rules which place tentative constraints upon future possible theories; these principles thus embody tentative knowledge about that of which we are as yet largely ignorant, but seek to know in greater detail.[5]

It was just this methodology of aims and methods being developed in harmony that was so successfully exploited by Einstein. The ruling passion of his life was precisely to know, to apprehend, to draw closer to, the magnificent, unified, intellectual fugue which he believed to be embodied in all natural phenomena. In developing his special and general theories of relativity, Einstein groped after ideas that might conceivably constitute the general principles, the unifying themes, of this conjectural fugue. He indulged in controlled metaphysical speculation concerning possible aims for theoretical physics, taking into account existing scientific knowledge, and taking into account the idea that some kind of unified harmonious pattern exists to be discovered. It was this that led him to postulate that the laws of nature ought to have the same form when referred to co-ordinate systems in constant relative motion to one another, since if no aether exists, then the question of which co-ordinate system is at rest, and which in uniform motion, must be a purely arbitrary terminological matter, a question of human convenience only, having no significance for the nature of reality. And this in turn led him to entertain the apparently extraordinary, paradoxical idea that the velocity of light ought to be the same with respect to all uniformly moving co-ordinate systems, since the constancy of the velocity of light is an apparently well-established physical law. The special theory of relativity is precisely the solution to the apparently insoluble problem of how the above two ideas can possibly be compatible with one another – two ideas proposed as a result of the concern to discover unity. And special relativity itself comes up with an invariance principle – known technically as Lorentz invariance – to which other theories of physics ought to conform. The principle of Lorentz invariance thus functions in physics as a methodological rule, of great further fruitfulness in physics, which places *restrictions* on the form that other theories in physics must take if they are to be acceptable.

The same kind of groping after unifying principles led Einstein to the development of general relativity. Einstein noticed that the effects of uniform acceleration, and the effects of being at rest in a uniform gravitational field, are precisely the same. Perhaps this is the clue to another unifying theme: natural phenomena do not distinguish between uniform acceleration, and gravitation. If so, we are led to a quite remarkable conclusion. For according to special relativity, acceleration has an effect on the measured geometry of space, and the measured flow of time.[6] Thus perhaps gravitation affects the measured geometry of space and the measured flow of time in just the same way. Perhaps, indeed, gravitation is just a kind of warp, a curvature, in the geometry of space-time. Here is a dazzling new idea, which seems to fuse together in an extraordinarily novel way the apparently distinct things of gravitation, the geometry of space, and time. Our aim in developing a new theory of gravitation, which has become necessary as a result of the advent of special relativity, should be to develop a theory which captures the above new, unifying idea, in as natural and simple a way as possible. The field equations of general relativity do precisely that: they amount to the simplest, most coherent way of capturing, in the form of a precise, testable theory, this new, unifying metaphysical idea that gravitation is a geometrical feature of space-time.

Thus these two quite extraordinary developments in our scientific knowledge embody exactly the kind of aim-oriented empiricist methodology that arises as a result of our modest tap applied to standard empiricism. Modern science at its most strikingly successful proceeds in accordance with our more rigorous ideal for scientific enquiry. The lesson to be learned from all this is surely that we should apply this new, more rigorous, amazingly successful methodology to the whole of science and technology, and not just to theoretical physics, in the somewhat half-hearted way we do at present, as a result of the impact left over from Einstein's profound, whole-hearted and instinctive use of this methodology.

SCIENTIST: Well, you have at least stimulated me to go off and have a look at some of Einstein's methodological writings.

PHILOSOPHER: That's marvellous!

CHAPTER SEVEN

THE IDEAL OF SCIENCE FOR PEOPLE

(Two or three weeks have passed. The Scientist has been avoiding the Philosopher, simply because the Philosopher has this terrible tendency to go on and on and on; each session is so time consuming! Finally, however, our Scientist decides to pay another call on his friend. Greetings have been exchanged.)

SCIENTIST: You will be pleased to hear that I have been having a look at some of Einstein's occasional writings.

PHILOSOPHER: Yes? What did you find?

SCIENTIST: Oh, I had a look at some of the things in *Ideas and Opinions*;[1] his *Autobiographical Notes* in the Schilpp volume.[2] His letters to Born about quantum mechanics.[3]

PHILOSOPHER: Don't you think he was a marvellous wise old bird?

SCIENTIST: Well, I certainly found some echoes of the things we were talking about – or echoes here of things Einstein wrote about. That marvellous passage on education, just thrown off casually in the *Notes*; the quite amazing depth of his passion for physics – his desire to draw closer to and comprehend, his "Old One". Very strange! It really was his whole life: the rest seems to have been somewhat incidental – except that, as a person, he seems to have been so full of life, sparkle and vitality …..

PHILOSOPHER: (interrupting) You know how I imagine it? It's probably completely stupid, but I think of the basic emotional impulse behind theoretical physics as a mood, a perception, a feeling, which is rather like something once expressed by D.H. Lawrence, casually tossed off in one of his letters. Let me see: where is the wretched book? (He hunts amongst his disorderly bookshelves.) Ah! Here it is. Listen!

"It is quite true what you say: the shore is absolutely primeval: those heavy black rocks, like solid darkness, and the heavy water like a sort of first twilight breaking against them, and not changing them. It is really like the first craggy breaking of dawn in the world, a sense of the primeval darkness just behind, before the Creation. That is a very great and comforting thing to feel, I think: after all this whirlwind of dust and grit and dirty paper of a modern Europe. I love to see those terrifying rocks, like solid lumps of the original darkness, quite impregnable: and then the ponderous, cold light of the sea foaming up: it is marvellous. It is not sunlight. Sunlight is really firelight. This cold light of the heavy sea is really the eternal light washing against the eternal darkness, a terrific abstraction, far beyond all life, which is merely of the sun, warm. And it does one's soul good to escape from the ugly triviality of life into this clash of two infinities one upon the other, cold and eternal." [4]

There! That to me expresses something of the *feel* of the thing. The *otherness*. The not-us. The utterly beyond-all-human-affairs, wholly and forever untouched by human affairs, here before us and here when we have all gone. The ultimate mystery.

(The Scientist laughs but does not commit himself.)

Einstein of course himself put it slightly differently. For example, listen to this.

".....there is a third stage of religious experience which belongs to all of them, even though it is rarely found in a pure form: I shall call it cosmic religious feeling. It is very difficult to elucidate this feeling to anyone who is entirely without it, especially as there is no anthropomorphic conception of God corresponding to it.

"The individual feels the futility of human desires and aims and the sublimity and marvellous order which reveal themselves both in nature and in the world of thought. Individual existence impresses him as a sort of prison and he wants to experience the universe as a single significant whole. The beginnings of cosmic religious feeling already appear at an early stage of development, e.g. in many of the Psalms of David and in some of the Prophets. Buddhism, as we have learned especially from the wonderful writings of Schopenhauer, contains a much stronger element of this.

"…..I maintain that the cosmic religious feeling is the strongest and noblest motive for scientific research. Only those who realise the immense efforts and, above all, the devotion without which pioneer work in theoretical science cannot be achieved are able to grasp the strength of the emotion out of which alone such work, remote as it is from the immediate realities of life, can issue. What a deep conviction of the rationality of the universe and what a yearning to understand, were it but a feeble reflection of the mind revealed in this world, Kepler and Newton must have had to enable them to spend years of solitary labor in disentangling the principles of celestial mechanics! Those whose acquaintance with scientific research is derived chiefly from its practical results easily develop a completely false notion of the mentality of the men who, surrounded by a skeptical world, have shown the way to kindred spirits scattered wide through the world and the centuries. Only one who has devoted his life to similar ends can have a vivid realization of what has inspired these men and given them the strength to remain true to their purpose in spite of countless failures. It is cosmic religious feeling that gives a man such strength. A contemporary has said, not unjustly, that in this materialistic age of ours the serious scientific workers are the only profoundly religious people." [5]

SCIENTIST: One point did occur to me as a result of reading Einstein's odd piece on methodology. Why didn't you present your aim-oriented empiricist ideal for science and technology as a *generalization* of Einsteinian methodology? The parallels seem to me to be quite close. Presenting your argument in this way would enormously strengthen it, especially in the eyes of scientists. Look what a case you can make out for your position: "Here is this magnificent episode in the history of physics: the development of special and general relativity, the birth of quantum mechanics, all brought about by Einstein, with his new methodological ideas playing a central role in the development of these theories. The time has come to generalize this Einsteinian methodology of discovery so that it becomes applicable to the whole of science and technology. Science at its most rigorous, and most strikingly successful, embodies not standard empiricism, but aim-oriented empiricism. The moment we adopt aim-oriented empiricism as the ideal of rigour for science we see that it is *essential*, if science is to be truly rigorous and objective, that non-scientists, as well as scientists, together try to determine the best aims for science. To this extent science at its finest and most successful, embodies principles close to your ideal of a people's science." You can even add: the methodology of aim-oriented empiricism can almost be selected on quasi empirical grounds, in that this methodology has already proved itself as strikingly successful. In essence, science as it already exists is highly rigorous: it is just that widespread, careless acceptance of standard empiricism has obscured this latent rigour from general view. I am not saying I agree with all this. Merely that it does seem to put your case across powerfully and succinctly, and in a way which is unlikely to get the backs up of the scientific community. Why not try putting it across like that?

PHILOSOPHER: (ruefully) It seems to me that your exposition is a distinct improvement on mine.

SCIENTIST: Then why didn't you present it to me in those terms?

PHILOSOPHER: I don't know. It may have had something to do with the way the idea first came to me – as a result of pondering the inadequacy of Popper's methodology. I was concerned with the problem of the rationality, the rigour, of science: Popper's viewpoint, it seemed to me, despite his own claims to the contrary, failed to solve the problem – the problem of induction. I began to realise that Popper failed to solve the problem because he – along with so many other philosophers and scientists – misrepresented the basic aims of science.[6] Science does not seek truth *per se*: science seeks explanatory truth, intelligible truth, beautiful truth, desirable truth – i.e. valuable and useful truth. The whole enterprise of science of course presupposes that valuable truth exists to be found: science, at its most rational and rigorous, rests on faith, faith in the value of human life, faith that the world can support and nourish human life. But just this "faith" – unless left in an extremely vague form – is highly problematic. Possibilities need to be articulated, explored, developed, scrutinized. The moment

115

one realises that metaphysical ideas dominate science (for example, the idea that Nature has a simple, unified mathematical structure, at least to a high degree of approximation), to such an extent that theories which clash too violently with those metaphysical ideas are never even formulated, let alone considered, whatever their empirical success might be if they *were* considered, it becomes clear that rigour demands that these immensely influential and highly problematic metaphysical conjectures need to be made thoroughly explicit, precisely so that they can be criticised, improved. The whole history of attempts to make rational sense of science – from Descartes, Locke, Newton, Hume, through Kant, Comte, Mill, down to our own times of Russell, Duhem, Poincaré, Popper and others – this whole history I saw as a history of attempts to conceal the vital point that metaphysical presuppositions are conjectural, and hence in need of constant revision and development, as science progresses. The empiricists denied the existence of *a priori* presuppositions; the rationalists claimed to be able to establish the truth of these presuppositions by reason; Kant claimed to be able to show that experience is only possible if our experience substantiates such presuppositions; the conventionalists argued that giving *a priori* preference to simple theories in no way involves committing science to the *a priori* presupposition that Nature herself is simple; the positivists, absurdly, denied the very meaningfulness of metaphysical ideas; and finally Popper argued that simplicity equals degree of testability, *a priori* preference for simple theories thus really being no more than wholly respectable preference for theories of a high degree of testability or falsifiability. All these people were trying to make rational sense of science. They were all concerned with scientific rigour. And yet precisely this concern had the consequence, if anything, of helping to sabotage scientific rigour. For instead of loudly proclaiming the one simple, crucial point that needs to be recognized if we are to have truly rigorous and successful science, namely, our basic metaphysical presuppositions about the world ("Nature is simple") are conjectural, and hence in constant need of scrutiny and improvement, as an integral part of science, they sought to achieve the *absolute opposite* of this. With great ingenuity, in all kinds of diverse ways, they sought to conceal this one simple point. These metaphysical conjectures (a) did not exist; (b) could be proven by reason; (c) could be known to apply to all experience with certainty; (d) were meaningless; (e) were in no way brought into science by scientists' *a priori* preference for simple theories. All that labour, all that sophistication, complexity and solemnity; and at the root of it all, such elementary idiocy.

The very attempt to make science rigorous was what prevented science from being truly rigorous![7]

It was, of course, clear to me that science *in practice* proceeded in accordance with aim-oriented empiricism – or it would not have made any headway at all. It was also clear to me that science could only proceed in accordance with aim-oriented empiricism in a somewhat awkward inefficient way precisely because scientists' intellectual conscience demanded that they ought really to be attempting to proceed in accordance with some variant of standard empiricism. But not for a moment did I think that any scientist would have openly and consciously *practised* aim-oriented empiricism; just because to do that would have flown too openly against our whole intellectual tradition. Then what did I find? First, it began to dawn on me that Kepler had done just that. Kepler, so often charged with "irrationality" by the stern moralistic guardians of science's rationalistic reputation, had in fact all the time been the supreme rationalist, the supremely honest and rigorous scientist. How my heart went out to Kepler – especially after reading Koestler on Kepler.[8] It then began to dawn on me that Einstein's contribution to physics represented an amazingly successful exploitation of aim-oriented empiricism – as you have pointed out. I rushed to read up Einstein's few brief essays on methodology, and there it all was! Oh, how my heart went out to Einstein. So patient, so good humoured, so kindly; so basically unperturbed by the fact that his colleagues had not really even begun to notice how he had transformed science, knit together again fragmented science and philosophy to form the real enterprise, Natural Philosophy.

SCIENTIST: Hum! You *may* be right in what you say, for all I know. But I am not really well enough informed about the history of the philosophy of science to be in a position to tell. For all I know, it could all just be your customary rampant megalomania. One point does, however, occur to me. If all these highly intelligent men really have been persistently missing this simple, elementary point that you wish to emphasize, how do you *explain* it? What blinded them to this elementary point?

PHILOSOPHER: Emotion! And to be more specific: Fear! Fear of reality.

SCIENTIST: *What?* I don't believe it!

PHILOSOPHER: I am serious! You know this idea I have that science is concerned with *relationships*: relationships between people and Cosmos, and people and people? Well, for two or three centuries, science has been trying to establish an utterly absurd relationship between Man and Cosmos – a relationship of Man *dominating* Nature, having power over Nature. Around the time of Newton, natural philosophers talked quite glibly of "torturing Nature to reveal her secrets". The Idea that natural philosophy would enable Man to acquire power over Nature, thus becoming more God-like, was a commonplace. "Knowledge is power", puts the thing in a nutshell.

Originally all this seemed to be entirely sanctioned by Christianity. God had constructed a lawful, rational universe, so that Man could acquire knowledge of these laws with his Mind, thus acquiring God-like power over Nature. Then, of course, science became detached from religion, and from the expression of feeling: scientists no longer spoke of "torturing" Nature: they simply continued with their experiments in a de-emotionalized way. The underlying emotional impulse to acquire power over Nature through knowledge remained however the same.

Given this basic emotional impulse behind science, it was absolutely impossible for those caught up in the scientific quest, in this spirit, to acknowledge the wholly conjectural character of all scientific knowledge – and especially the conjectural character of basic metaphysical presuppositions about the nature of the universe. For, of course, if knowledge is conjectural in character, then Man cannot possibly acquire Power over Nature through

Knowledge. However much we know, Nature may always, at any moment, surprise us utterly.

Just this was, of course, pointed out by Hume. And the response? Hume must be wrong! The scientists could dismiss Hume's arguments, because they belonged to foolish "philosophy", which never got anywhere anyway, whereas *science* was progressing in leaps and bounds. The philosophers struggled to refute Hume's elementary arguments; and, of course, failed in the attempt.

No one realised that Hume was absolutely *right!*[9] Hume has pinpointed the *intellectual* absurdity of the basic *emotional* absurdity underlying science. Of course we cannot put ourselves into a position of God-like power over Nature. The very idea that we can achieve such a relationship of power can only be a hysterical cover-up for extreme fear.

The attitude of Kepler and Einstein is quite, quite different (there are, of course, many others as well: I give here only a kind of cartoon sketch of the thing, in black and white, with no tones of grey). Kepler and Einstein are profoundly moved by the *beauty* of the universe – a beauty which they believe we have got odd glimpses of in our physical theories. There can, of course, be no question of knowing for certain that the beauty really is there; but they both put their trust in its existence; it is their faith that it exists. They adopt the attitude of trust and love. And with all the passion and sensitivity of love, they seek to see this beauty, divine it, in a highly speculative fashion, always ready to acknowledge that ideas dreamed up, despite their apparent appeal, may have no basis in reality. It is *reality* that they seek to know and love, not their fantasies of reality. They are like lovers, seeking to divine what is best in the beloved; or like music-lovers, seeking to divine the inner harmony and beauty of great music heard through static on a radio.

Just such an attitude of trust and love enabled Einstein to put his trust in the idea that Nature has within itself an inner unified mathematical harmony; as he, himself, has said, following up this 'speculative metaphysical idea' led him to develop the special and general theories of relativity. General relativity in particular could never have been discovered, or created, without Einstein's passionate conjecture that the universe has an inherent harmony to it, and without the rational *use* of this idea. The general theory of relativity is one of the most beautiful songs that have been sung to the universe, to "make the universe happy" as a Pygmy scientist would say. Einstein did not *fear* the world, as a result, absurdly, seeking to gain power over the world: he loved the world; and he put his trust in its lovability. As a result of seeking beauty, he found it – or so we can believe, if we wish.

SCIENTIST: Well, well, well. So this is what Pygmy science is all about. Science is our expression of our love for our world – at its rational best. If based on fear, on the lust for power, it becomes irrational, incapable of being rationally understood – as the insolubility of the problem of induction, as traditionally conceived, indicates.

PHILOSOPHER: You've got it exactly.

SCIENTIST: Well I'm glad I have at least *understood* your viewpoint. And now I'm afraid I must dash, as –

PHILOSOPHER: But we have only just begun!

SCIENTIST: We've only just begun?

PHILOSOPHER: Yes! So far, we have made only the first tiny step. Almost everything of importance is to follow.

SCIENTIST: Oh! (He glances with resignation at his watch.)

PHILOSOPHER: (Excited, not noticing, or choosing to ignore it if he has noticed) In fact, that is why I did not employ your exposition! Because I have my eye on the whole argument. Our first step might be put like this. We begin with the ideal of standard empiricism: the basic aim of science is to discover value-neutral factual truth. Tap! A slight enhancement of rigour, and we have aim-oriented empiricism: the basic aim of science is to discover valuable truth. At once it becomes clear that the basic aims of science are profoundly problematic. If science is to be truly rigorous, truly objective, scientists and non-scientists alike must collaborate in seeking to discover the best possible aims for science. The needs, desires, feelings, problems, aspirations, ideals

120

of *people* must find expression and representation within the intellectual domain of science, on the level of aim articulation and aim exploration, instead of being ruthlessly excluded from the intellectual domain of science, as standard empiricism demands, in the name of "rigour".

This slight enhancement of rigour has brought with it both intellectual and human riches. We have something like a *rational*, though of course fallible and non-mechanical, method of discovery in science, a method which enhances our capacity to point our noses in the direction of valuable truth guessed to exist, but as yet undiscovered. At the same time, we have the hope of developing a science that is more intelligently and sensitively responsive to human social needs, problems, desires, aspirations, ideals.

We are one step nearer to our ideal of a truly rigorous person-centred science. Personal feelings, desires, problems, ideas, have a rational role to play within science, on the level of aims. It ought to be possible for "any old fool", as you put it, to win a Nobel prize for contributions to science. The intellectual domain of science is not entirely *dissociated* from human life.

On the other hand, at the level of theory and experimental and observational data, we still have the intellectual domain of science dissociated from the rest of human life, as required by standard empiricism. Our next step, our next enhancement of rigour, changes all that!

This time we begin with aim-oriented empiricism: the basic aim of science is to discover valuable truth. And we ask the simple question: Why? Why do we seek valuable truth? What more distant, or more general, goal do we have in mind in seeking to discover valuable truth? Is the discovery of valuable truth an end in itself; or is it a means, a stepping stone, to some broader, more distant end?

The answer is surely extremely obvious! We seek to improve our knowledge of valuable truth ultimately in order to make such knowledge available to people to be of use and of value in their lives. The ultimate aim of scientific enquiry is to be of help, of use, to people in their lives. Our ultimate concern, as scientists, is to help people emerge from poverty, illness, misery, semi-starvation, grimly restrictive living conditions, into a life of health, prosperity, happiness, freedom, fulfilment. Ultimately, and ideally, the concern of science is to be of assistance to people in their search, their struggle, to create and live a richer and more rewarding way of life. The basic aim of science, we may say, is to help promote human welfare, help enhance the quality of human life. Our concern to discover valuable truth is but a means to this end. Ultimately, what matters is human life: and scientific knowledge and understanding matter only insofar as they are of use and of value, of practical service, in helping us to enhance the quality of our lives, in helping us to realise that which is of greatest potential value in our lives.

This, then is our second step. Science does not seek valuable truth *as such*; rather it seeks to be of service to people in their lives: ultimately valuable truth is sought solely as a means to this end. The basic aims of science need, m short, to be conceived of in human, personal, social terms. The intellectual progress of science needs to be assessed in terms of the potentiality of science to help promote human, personal, social progress.

SCIENTIST: All very fine. I don't think I would really want to quarrel with any of this. But why does all this constitute an enhancement of rigour, of objectivity? Isn't it a rather commonplace idea that the whole *raison d'etre* of science is to help promote human welfare?

PHILOSOPHER: Yes, it is, I think, a commonplace idea. It is certainly a very *old* idea. It was this idea that inspired Bacon, for example. My point, however, is that it is not really taken seriously within the context of scientific research. Or rather, because of the widespread attempt to make science conform to standard empiricism, science has not pursued its basic aim of helping to promote human welfare in a very intelligent, effective, rational and objective fashion.

The thing might be put like this. If, in pursuing some activity we seek to realise a goal T, solely as a means to the realization of the more distant goal, H, and we become so obsessed with our concern to realise T that, in pursuing T we forget entirely *why* we are seeking to realise this goal, then clearly our activity has become profoundly irrational, indeed almost lunatic. We will neglect to use our attainment of T to attain H; we will neglect to tackle the problems that arise in using our attainment of T to attain H; we will fail to consider the possibility that, in order to achieve our basic aim, H, it might be better to seek some slightly modified aim T* as a stepping stone to H, rather than T. Construed as attempts to realise our basic goal H, our T-seeking activities will have become hopelessly inefficient and irrational.

My claim is that all these rationalistic failings can be discerned in science as it exists at the moment, and indeed are almost inevitable as long scientific enquiry takes as its immediate aim the discovery of valuable truth or knowledge (T) rather than the discovery of valuable truth or knowledge *in order* to help promote human welfare (H). Scientists do indeed tend to suppose that their professional scientific task is at an end when they have discovered valuable truth, and have published their findings in the appropriate technical journal; they tend to forget to ask the further crucial questions: For *whom* is this truth of value? Have I communicated my results to those who most need it, or might value it? Have I expressed myself in a way that is comprehensible to those who might most need or value this new knowledge?

Again, scientists do tend to neglect the problems that arise in connection with the human *use* of scientific knowledge. Immense care, attention, skill, intelligence is devoted to the problems of acquiring knowledge; a kind of vast indifference and neglect arises, however, over problems that arise in connection with knowledge being understood, valued and *used* by people in their lives.

But the real trouble in my view, is precisely the dissociation of the search for new knowledge from the search for a better way of life, the realization of desirable, personal, social ends. My claim is that scientific, intellectual problems need to be understood as an aspect of personal, social problems. Conceiving of scientific, intellectual problems in this way, would both enhance our *understanding* of scientific, intellectual problems, and would enhance our capacity to develop a science sensitively responsive to personal, social needs and aspirations. The intellectual aspect of science would be enhanced; and the personal, social aspect of science would be enhanced as well. In short, it is precisely the dissociation of the intellectual domain of science from the personal and the social, even at the level of theory and experimental and observational results, which renders science inefficient, unintelligent or, in other words, somewhat irrational or unrigorous when science is conceived of as seeking fundamentally to help promote human welfare.

SCIENTIST: I am not sure that I have the faintest idea what you are talking about.

PHILOSOPHER: Let me put it in terms of a diagram. In essence, the points that I want to make are extremely simple and obvious. They constitute obvious developments of our previous points. They only seem at first sight to be somewhat abstruse because they run so counter to the whole way in which science has traditionally been conceived of. The diagram is not very good, I am afraid, but it's the best that I have been able to think up. (He draws diagram 4.)

In terms of the diagram, arrow C, the activity of bringing new scientific knowledge and understanding to people who might need it or value it is at present neglected. The problems posed by this activity are ignored. And more fundamentally, the route B + C (aim-oriented empiricist science + the activity of making scientific discoveries available to people) is a more roundabout, less efficient and intelligent a pursuit of the goal of helping to promote human welfare than the route D, which represents the Ideal of Science for People. This third direct route clearly represents to itself within the context of scientific enquiry, that the aim of science is to be of help, of value, to people. Thus scientific, intellectual problems are conceived of as an aspect of personal, social problems: scientific knowledge is conceived of in personal, social terms. Just this, I claim, would enhance our understanding of scientific, intellectual problems, and would enhance our capacity to develop a science capable of sensitively and intelligently helping to promote human welfare.

SCIENTIST: Can you give me some examples of all this? I still hardly know what you are talking about, I am afraid. I am not yet even in a position to *disagree* with you, because I don't yet understand you.

PHILOSOPHER: Yes, of course. Let me see: How shall we proceed? Shall we begin by considering technological science?

SCIENTIST: Fine.

PHILOSOPHER: In the case of technological science, the thing is surely especially easy to see. According to the ideal of *humane* aim-oriented empiricism – the ideal of science for people – the basic aim of technological science is to help people solve their problems, realise their desirable human ends, by developing appropriate, useful technology, technology sensitively adapted to the requirements, the needs, the capacities of people. It is surely clear that science can only hope properly to fulfil this task if expert, technical problems are clearly understood to be but an aspect of the *human* problems, the personal, social problems. It is only if the *human* problems are put first that technical problems can be chosen and tackled in such a way as to be sensitively appropriate to solving human problems. If technical, scientific problems are dissociated from the human problems they are designed to solve, then the great danger is that technical problems will be tackled and solved because of their inherent intellectual interest, because they have suddenly become possible to solve because of recent developments in knowledge, or because money is available, from some source or other, to develop solutions to just these problems. The question of whether solutions to these technical problems actually help to solve human problems, in genuinely beneficial ways, or whether new human problems are actually being *created* by these new technological solutions, will tend to be disregarded. It will be "unscientific" to consider such issues, beyond the province of the expert.

Pollution, depletion of natural resources, modern armaments, dangerous chemicals in our human environment – elementary and horrifying things like lead in petrol – are all examples of

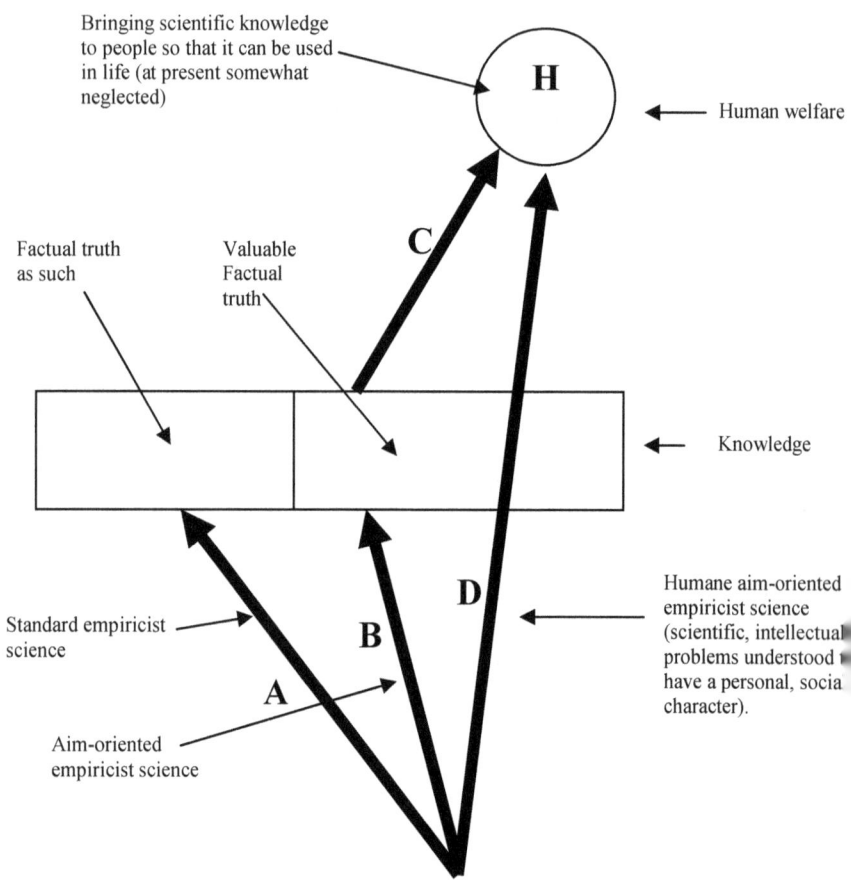

Bringing scientific knowledge to people so that it can be used in life (at present somewhat neglected)

H

Human welfare

Factual truth as such

Valuable Factual truth

C

Knowledge

Standard empiricist science

Aim-oriented empiricist science

B

D

A

Humane aim-oriented empiricist science (scientific, intellectual problems understood have a personal, social character).

DIAGRAM 4

technology developed without sufficient attention being paid to human consequences. One may well feel too that brilliant technological exploits such as putting men on the moon, or the development of Concorde, are examples of technological problems solved without sufficient attention being paid to elementary human needs and priorities. Elementary and horrifying facts about our world seem just to be ignored – such as the fact that millions of children suffer permanent brain damage during the first few years of life simply because they do not get enough to eat, their full personal potential lost for ever.

The tendency to dissociate technical, intellectual problems from human problems will have a bad effect on those who are responsible for making available scientific technology to people in order to be of help in solving human problems. Thus, those who work in hospitals, for example, will tend, as a result of their scientific education, to see technical, scientific problems and not human problems. Illness will tend to be conceived of by doctors as a technical, biological, medical problem, and not as a human problem with a technical, medical aspect. One will not tend to find hospitals run on the assumption that the whole purpose of the hospital is to render available to the ill person all the technical facilities of the hospital so that *he* can, insofar as he is able, use them in order to recover his health. The attitude will not tend to be that it is the *patient* who should make the decisions, direct operations, as it were, insofar as he can, doctors, nurses and surgeons simply helping out with their specialized facilities and capacities. This will not tend to happen just because scientific medical training will tend to emphasize medical problems divorced from the human problems of which they are a part. Medical experts will tend to come to see patients in rather the same terms as the TV repair man sees malfunctioning television sets: there is a technical fault here which needs to be put right, and the odd noise (of desperate anxiety) that the object of our attention is emitting is probably no more than a further sign of technical malfunctioning, which can easily be put right by technical means – administration of tranquilizing drugs in the case of patients, pulling out the plug, in the case of television sets.

You must understand: I am not accusing doctors of inhumanity, or anything like that. I am simply describing tendencies that are *bound* to arise as a result of conceiving of, practising, and teaching science in accordance with a seriously unrigorous ideal for science – an ideal which dissociates scientific, intellectual problems from human problems, instead of emphasizing that scientific, intellectual problems are an *aspect* of human problems, an *aspect* of our attempts to solve human problems.

Indeed, members of the medical profession are themselves just as much victims as the rest of us. So many medical students must take up medicine to a considerable extent out of a desire to be of help, of use, to others: and what do they find? The human problems, which they wanted to help solve, have been lost amongst a mass of purely technical, scientific, intellectual problems. All too rarely will scientific, medical problems be put into their human context. It will be very difficult for a student subjected to this kind of education (administered from the very highest of motives) to keep alive his attention, his sympathies, for the human problems, medical problems being understood and appreciated as a part of that central concern. There will be no encouragement to do this, and every discouragement. And God knows, the problems that doctors themselves have to face, in treating their patients as people are serious and difficult enough as it is. It means coming to grips with the personal problem of death, on a personal level – something which it seems to me our society, our culture, does not know how to do very well.

All this, of course, has a much broader relevance than just to medical science. In my own teaching, for example, I am continually coming across science undergraduates who originally took up their scientific subject from the very finest of humanitarian motivations: and they find that such motivations are completely and utterly irrelevant to their scientific *training* – for that is what their "education" so-called amounts to. At no point are scientific intellectual problems rationally related to human problems and aspirations, to personal, social problems and needs. Personal interest in intellectual problems is not excited and stimulated by a rigorous presentation of these problems as arising out of our concern with the problems of living. Science, it comes to seem to most students, is only very distantly and tenuously connected with *life*; mainly it is just something pursued for its own sake, or for *academic* reasons. The human point of it all but disappears from view. And, as a result, instead of stimulating interest, enthusiasm, excitement, most science courses, at both school and university levels, achieve, for most pupils and students exactly the opposite effects – precisely because of the unrigorous way in which science is presented and taught. The only way in which scientists themselves are able to understand and respond to the situation is to suppose that they will have to make their courses a little less "rigorous" by adding in some extra "human interest". Naturally, it does not occur to them that it is precisely the *lack* of rigour that is implicit in their teaching that is the cause of the trouble, students, in some respects, often having a better grasp of the situation than their mentors. You can see the trouble: scientists are quite prepared to acknowledge that they are not brilliant teachers, inspired leaders of men and women, full of compassionate concern for the suffering of humanity. What they are not prepared to admit, and will find it very hard to admit, is just what they most need to admit in order to put things right, namely: their stupidity. The lack of rigour in their teaching, their whole way of conceiving of and doing science. It will not be easy, for example, for science teachers to ask students for their help: and yet clearly just this is what is needed. Education should, in any case, always be a two-way exchange of ideas and information. If teachers don't ask students for their ideas, their problems, their reactions, how can they know who they are speaking to, what effect their words are having?

Science teachers are worried about the flight from science. Perhaps they should be even more worried about the lack of rigour, the lack of rationality, inherent in so much contemporary science. Put that right, and the truly wonderful thing that science itself is will shine forth and will attract all the people that anyone could ask for. Many young people today are only too anxious to find some way in life of making a contribution that is of genuine human value – if only this desire will not be exploited and perverted. They are put off science because they cannot see how it relates to life in a positive and good way. Science seems to them – with good reason

– to be only rather distantly and tangentially related to life. Exhibit science and, above all *do* science, in a truly rigorous way, as a rational response to human need, human problems, human aspirations, and people will come flocking to science. Science doesn't need to be tarted up, given artificial human interest and appeal: she needs to be understood and appreciated for the truly marvellous human creation, full of delight and compassion, that she is.

Actually, now I come to think of it, the situation is really far worse than all this might suggest. My claim is that if science is to be taught in a truly rigorous fashion, then the centre of attention needs to be *human* problems, personal, social problems, personal, social aspirations – scientific, intellectual problems being presented, discussed and understood as arising out of our attempts to help solve human problems, help realise desirable human ends. (This, of course, is the conception of science represented by arrow D in diagram 4) Such an approach to science education is almost unheard of. One does of course, occasionally find that some additional lectures are given on "social significance" of science and technology as a part of science undergraduate courses. This is, of course, a step in the right direction. But it hardly amounts to presenting scientific intellectual problems as arising out of our attempts to solve human problems.

As far as most science university courses are at present concerned, however, the situation is far worse than this would suggest. It is not just that the centre of attention is on scientific, intellectual problems, without these in any way being rationally related to the problems of life. Far, far worse, even scientific, intellectual problems are ignored, and instead *information* is imparted, *skills* are acquired. It is not just that students themselves never get an opportunity to articulate, explore, discuss their own intellectual-personal problems, so that, as a result, they lose sight of them, and real education is submerged beneath the tedium and insult of training and indoctrination. Worse even than this, many science courses never even articulate the problems that led to the development of the scientific knowledge that we do have. Scientific problems, solved and unsolved, are simply never mentioned – apart, of course, from technical exercises in problem solving, problems with known answers handed down to students to check that they have mastered the intellectual tricks that they have been trained to perform. Problem-solving lies at the heart of science, at the heart of all intellectual activity. Science, properly understood and properly used, is simply an invitation and an aid to the articulation, exploration and resolution of one's own problems. This is the *essential* thing one needs to experience, to understand, in order to understand and appreciate science. And yet students are rarely, if ever, given an opportunity to discover this wonderful human use of science!

Simply in order to understand a theory, a result, one needs to understand the problem it was designed to solve, the purpose it was designed to help realise. Often science courses make no mention whatsoever of these kind of background problems – or give only a very distorted picture of what the problems really were.

Again, in coming to understand a piece of science, a new idea, theory or result, we may well need to solve a number of *personal* intellectual problems. "This cannot be right, because ..... Why should it be done just like that, and not in this way, or this way? What is the point of it, the purpose behind it? Why is this idea significant? Why should I accept it?" These are the kind of objections and questions that should come tumbling from our lips when we are presented with a new idea, a new theory, result or technique. It is just our capacity to ask such questions that represents our *intelligence*, our *creativity*; it is precisely as a result of asking such questions and discovering to our own satisfaction that acceptable answers exist – at least up to a point – that we can make the idea our own, that we can freely choose it for ourselves (as opposed to having it inserted into our heads by the educational situation). And what happens in practice? Our need to raise such objections, difficulties, problems, represents our *inability to understand*, our *stupidity*,[10] just that which education is trying to eliminate. The very thing that should be nourished, stimulated, encouraged to grow and flourish, namely our intelligence and creativity, our independence of mind, our intellectual integrity, becomes the one thing that education is seeking to eradicate, to eliminate, to annihilate. To this extent, formal academic education, and in particular science education, the worst offender in this respect, amounts to a massive and profound insult to the pupil, the student. It is as if the student is told, in the whole way in which the indoctrination process proceeds: "Who do you think you are? You don't know enough yet to have interesting important questions and problems. *Your* questions and problems are just indications of your stupidity, your ignorance. Shut up! We will tell you what questions to ask. And if you are very good, and learn up all that we have to tell you, and if you manage to perform the tricks that we set you, in order to check up on whether you really have absorbed it all, then, at the end of this long process of total humiliation, we will allow you, guardedly, to ask a few appropriate questions. But only when you have obtained your Ph.D. will you really be in a position to do research."

Is it to be wondered that students grow bored with their subjects? In the circumstances, what greater sign of intellectual health could there be than an attitude of indifference or hostility towards science?

And yet this educational horror story, acted out day after day in our schools and universities, does not even begin to touch the point that I was making above. As an elementary act of educational good practice we need to put scientific intellectual problems at the centre of attention, scientific results and theories being appraised from the standpoint of their capacity to solve problems. We can then perhaps go on to relate scientific, intellectual problems in a rational way to human problems, to problems of living.

Why is science teaching by and large in such a deplorable state? It is yet another example of the lack of rigour implicit in standard empiricism or aim-oriented empiricism. Scientists are so busy seeking to attain knowledge (arrows A and B of diagram 4) that they forget all about the problems connected with rendering this knowledge available, accessible and enjoyable to *people* (arrow C).

SCIENTIST: At last I am beginning to appreciate the ….. what shall I call it? Yes! ….. the *nuances* of your style. When you speak of "educational horrors" you mean you have a suggestion as to how things might be improved a little. (The Philosopher laughs!) I can see the trouble. You have a suggestion to make. No one hears you make it. So you raise your voice, in order to attract attention. You angrily denounce everyone right, left and centre. You begin to scream. And, of course, as a result, no one listens! The world is full of madmen angrily denouncing everyone right, left and centre. Your voice breaks, you fall silent, and you conclude that perhaps you have, after all, got everything a little out of proportion. If there is a madness sanity mismatch between you and the world, then it's got to be *you* that's mad. Given the choice: either the world is mad and you are sane, or the world is sane and you are mad, then, by definition, as it were, you are mad. Isn't that how it is?

PHILOSOPHER: (Grinning) How perceptive you are. You understand my situation exactly! Not that I really have any doubts about the solution. Sanity consists in learning to cope in an enjoyable and fruitful way with the world's insanity, never losing sight of the essential point that most people are both sane and good. There: how does that sound?

SCIENTIST: Very sane.

PHILOSOPHER: But before we leave the subject of education, let me tell you a little story – a true story, incidentally. It happened to a friend of mine. He, too, is a philosopher who goes on and on about the aims of science, the human meaning of it all, etc. He teaches a few science undergraduates in his College. One day, he went along to the Physics department, to try to persuade them to release the odd interested physics undergraduate, so that they could do his course – one hour a week for one academic year! They were interested, simply because they were worried about falling student numbers, and they thought "philosophy of science" on the syllabus as a possible option might boost the image of the course, in student eyes. And then the subject of quantum mechanics came up, and the problems of knowing how to interpret the theory. The following conversation then took place between the departmental tutor to the physics undergraduates and my friend.

Physics Tutor: But you wouldn't of course discuss the problems of quantum mechanics while they were still learning quantum mechanics!

Friend: (After a slight hesitation) Yes. Why not?

Physics Tutor: But what would you say? Would you solve all the problems? Or would you leave them all wide open?

Friend: Well, hopefully, I would do something in between these two extremes.

Physics Tutor: (Speechless at the educational enormity of discussing real intellectual problems with undergraduates).

I might add that the College evidently felt that this Tutor had such a good grip of educational matters, that it appointed him to act as liaison officer to schools, to encourage school leavers to take up science.[11] So you can see. It goes on, it goes on. For example, just think what goes on in the average mathematics lecture at most universities! Lecturer copies notes onto blackboard, mumbling away. Students copy hieroglyphics from blackboard into their notebooks. All these highly intelligent people gathering together day after day to turn themselves into copying machines. Discuss the problem? Raise questions about aims, motivations, the human point of it all? Consider the way the thing developed? Explore difficulties, objections, criticisms raised by students? Work towards getting an intuitive feel for things? What rubbish! Leave the lecture-room at once! Stop disrupting the orderly process of education.

Life, the world, is so full of riches, so full of extraordinary, beautiful, dramatic, moving, painful, difficult, tragic things. Science is there to help us to discover, to see, to experience; and to help us put things right, make things better. And people never get a chance to discover for themselves this wonderful human use of science: for whenever they come across science, in education, it seems to consist only of being stuffed full of some nasty tasting, anti-life substance. And those who hand it out, do it with such a sense of duty and responsibility.

SCIENTIST: But –

PHILOSOPHER: I have just three more tiny, miniscule points to make, and I promise, we will be done with this part of the argument.

First point. The idea that the scientist's main aim is to improve knowledge (arrow A or B of diagram 4), the further question of for whom this knowledge is intended, and why, for what purpose, being largely ignored, has the consequence – as I have, I think, already mentioned – that scientists tend not to regard it to be a part of their *professional* obligation as scientists to ask questions such as: who needs this knowledge I have developed? How can I tell them about it? How can I put it in a form which can be understood and used? On the contrary, scientists tend to assume that if they have published it in the appropriate technical journal, their professional obligations, as research scientists are at an end. The rest, surely, will take care of itself, won't it?

Second point. The same idea leads to what is, in my view, a misplaced emphasis when it comes to scientific, intellectual values and priorities. Once we adopt the idea that the whole point of science is to help promote human welfare, then it is clear that science only achieves its basic objectives when the products of scientific research come into contact with *people*, and are used by people to enrich their lives. A scientist who gets a new idea for a piece of valuable new technology helps to make something possible: he thus makes a potential contribution to science. But it is the engineer, the technologist, the chemist or whoever, who develops the original idea, makes it practicable, brings it to life, to reality, who makes the real contribution. And perhaps even further, it is those who actually manufacture and render available to people the product who bring things to fruition, as it were.

The idea that the main aim of science is to acquire knowledge once again distorts all this. It leads to the conclusion that only the person who makes the *theoretical* contribution really makes a contribution to science: all the rest simply work out some rather messy, practical, scientifically uninteresting details. Just when science is about to reach fruition – that is when it is about to come fruitfully into contact with other people, with life – scientists lose all interest. No Nobel prizes are to be won in the field of

135

technological development – just because such work is not conceived of as making contributions to *knowledge* and hence to science. Inevitably, science being conceived of in this way, the finest scientific minds will tend to restrict their attention to pure research, to making a contribution to science, leaving to others the relatively mundane task of applying new knowledge to help solve the problems of life.

And all this means, of course, that the centre of intellectual attention will not be human problems, social problems, technical problems needing to be sensitively inter-related with the feelings, problems, desires, ideals of people, if the most humanly valuable kind of technology is to be developed.

You can see the trouble. In every case, these problems arise because scientists, in practice, in pursuing scientific research, take as their main aim the improvement of knowledge (arrows A or B). They forget to ask: who needs this knowledge? For what? What are people's needs, problems? (arrow C) And they certainly do not take as the central active concern and aim of their scientific work: to help promote human welfare, to help solve human problems (arrow D).

SCIENTIST: There is, I think, something in what you say. But what is to be done? Scientists are only human after all. They are not saints.

PHILOSOPHER: But it is not a question of sainthood at all. It's simply a question of making science a little more *rigorous*. The fundamental aim of science is to help promote human welfare. This is serious: not just waffle. Acquiring new knowledge is but a means to this end. Doctors after all manage in their professional capacities to help promote human welfare, without necessarily being saints. Are we not entitled to expect of scientists what we automatically expect of doctors, in their professional capacity? Not for a moment do I think that the concern to help promote human welfare is not present, within the scientific community. It is just that this very real concern is not at present being very intelligently pursued. That is my whole point. Scientists really do want to help promote human welfare. It's just that their unthinking acceptance

136

of a deplorable ideal of scientific rigour is actually disrupting the intelligent, effective, sensitive pursuit of this aim. Standard empiricism, and to a lesser extent aim-oriented empiricism, dissociate scientific, intellectual problems from human, social problems, so that scientists, in pursuing their research, inevitably find it very difficult to *understand* their scientific intellectual problems as an aspect of human social problems. It is just that which needs to be changed.

SCIENTIST: Yes, but how?

PHILOSOPHER: Essentially by taking seriously, and putting into practice the ideal of scientific rigour of humane aim-oriented empiricism. Scientific progress is to be assessed in terms of human progress. People who bring scientific discoveries in a fruitful way into life make just as important a contribution to scientific progress as those who contribute to "pure" research: such people thus deserve to obtain the rewards, the prizes, the honour, that goes to those who make important contributions to science. Such developmental work needs to be recognized as *scientifically* important if it has genuine human value. Again, if scientists are to pursue their research in a truly rigorous fashion, then they need constantly to ask themselves: Why? Why? Why? Why am I doing what I am doing? What is the human point of it? How might it conceivably be of human use or value? Is there some other research that I might be able to do of greater possible human value? In what ways ought my discipline to be of benefit to people? Are my fellow scientists concerning themselves with all the different ways in which our discipline might be of value to people? What ideas are there around in society as to where our most urgent personal, social problems lie? How can I succeed in interesting non-specialists in my specialist work? What reactions do non-specialists have to my work? Is there anything of importance to be learned from their ignorant reactions, their ignorant criticisms?

And, of course, scientists need *help* with all this from non-scientists. It is just we non-specialists, in no way entangled in the thickets of specialized, academic knowledge, who may be able to see most clearly and simply the extent to which a science, a piece of research, succeeds or fails to be of real value to people. We are people, after all. And so, of course, are specialist scientists. It's just that traditional bad ideas of scientific rigour have tended to suggest to scientists that in their professional work they ought to be depersonalized *intellects*, rather than *people* with feelings, desires, values, ideals, all of relevance to their specialized scientific thinking. We non-specialists can help to remind specialists that they are people even amongst the thickets of their apparatuses, jargon and hieroglyphics.

SCIENTIST: Surely you would want to recognize that many scientists today are concerned to develop their research in humanly fruitful directions?

PHILOSOPHER: Yes, of course, I –

SCIENTIST: And is there not, after all, the *British Society for Social Responsibility in Science* (BSSRS), concerned actively to campaign for some of the things which you are talking about here, and actually putting out a journal called *Science for the People?*[12]

PHILOSOPHER: Yes, of course. That is my point. There really is a powerful concern, within the scientific community to develop science in directions genuinely helpful to humanity. My essential point is almost quite trivial, and can be put like this. Scientists have failed to recognize that their *desire* to develop a more humanly fruitful science is constantly being hampered by unrigorous conceptions of science unthinkingly accepted. Human desirability, and requirements of scientific rigour only seem to be at odds with one another because customary concepts of scientific rigour are so unrigorous. Develop more rigorous ideals for science, put them into practice, and our very ideals of rigour will actively assist us in developing science in humanly fruitful directions, rather than obstructing us.

That is the point that I think is at present being missed. Members of BSSRS, for example, tend to locate the source of the trouble within Capitalism. My point is that our thinking, all our thinking, scientific and non-scientific, has been somewhat perverted by widely held irrational ideals for reason and rigour. The first thing we need to do is to put our thinking straight. We can then go on to think about the question: To what extent is the source of the trouble to be located within the capitalist system? At present we cannot declare "person-centred science is impossible within the capitalist system" for the simple reason that it has never been *tried*. The *idea* has not really been around.

SCIENTIST: May I make a general comment about your argument so far?

PHILOSOPHER: (Grinning) I think that would be permissible.

SCIENTIST: You have I think made out a fairly convincing case for saying that *technological* problems need to be conceived as an aspect of, a part of, personal, social problems, those faced by *people* in their lives. It is only if we *bring together* the feelings, desires, experiences, anxieties, frustrations, ideals of *people*, on the one hand, and the objective, abstract, technical considerations and problems of developing new technology on the other – it is only if we bring these two things together, and sensitively interrelate them, that we will be able to develop technology which suits our human requirements, and which leads, as you would put it, to the flowering of human life. The great danger of dissociating technical, technological thinking, from the desires, feelings and aspirations of people, is that technological development will proceed on its way, carried forward by the momentum of merely technical ideas and the profit motive, and we human beings will have to adapt *ourselves* to the monstrous technological world that we create, instead of adapting technology to suit our purposes, our needs and desires. All this I can agree with, as I am sure many others can as well. What I cannot agree with .....

PHILOSOPHER: Magnificent! As usual, you manage to express what I am trying to say much more coherently and succinctly than anything I seem to be able to manage myself.

Incidentally, one powerful and beautiful demonstration of just how important all this is in the medical field is provided by a fascinating book I have read recently, entitled *Awakenings*.[13] The author, Oliver Sacks, describes what happens when people who have had sleeping sickness take L-dopa. He argues for, and *demonstrates*, beautifully and movingly, that the central thing is the experience, the problems, the lives, of his patients: the consequences of neurological disorders, and administration of L-dopa, are profoundly affected by the human situation. Do read it. Despite some of the grim events he describes, it is in essence a profoundly heartening book, and powerfully illustrates the points that we have been discussing.

SCIENTIST: Yes, certainly. It sounds interesting. But you cut me off as I was about to offer you a criticism, an objection.

PHILOSOPHER: Oh. I'm sorry.

SCIENTIST: My objection amounts to this. The ideal of science for people may be very fine and good as far as *technological* science is concerned. After all, the whole idea of technological science ought to be precisely to develop products that are of use, of value, to people. The situation is however quite different, it seems to me, when we come to pure science, pure research.

PHILOSOPHER: (unable to contain himself) But, but, but …..

SCIENTIST: No please, let me finish. I know of course that you won't like this. We have already had one explosion on this subject. Nevertheless, it does just seem to me that pure science has a value of its own, independent of its value from the standpoint of technological applications. Scientific knowledge is worth obtaining and developing for its own sake; it is of value in its own right. This is why I feel that this new ideal for science you are proposing – the ideal of humane aim-oriented empiricism, or science for people – cannot encompass the whole of science. It has a certain relevance for technological science perhaps; but none for pure science. *There*. What's your reaction to that?

PHILOSOPHER: To be frank, discouragement. I thought we were getting on so well. And now I discover that there has been an almost complete breakdown of communication. I thought we were standing next to each other: now I see that a vast gulf separates our two positions.

SCIENTIST: Well, never mind. Gather together your courage, and tell me, simply and clearly, why *you* think what I have just said is inadequate.

140

PHILOSOPHER: Right. I will do my best. Let me begin by stating again, briefly, the central points of my position. It can be put like this.

All science, without exception, is of value insofar as it is of human value, of value for *human* life. Science is of value in two rather different ways (though in the end these two ways are but two aspects of one central thing: our relationships between ourselves, and between ourselves and the cosmos). Science is of utilitarian, technological or practical value, in that it helps us to solve our personal social problems, realise our desirable human desires, via technological applications of knowledge. And science is of *cultural* value, in that it helps us to improve our knowledge, our understanding, our appreciation of the world around us – encourages us to adopt a more perceptive, honest, objective, trusting, loving attitude towards the world about us. All science is applied science, applied, in one way or another, to helping us achieve fulfilment in our lives. The distinction between pure and applied science needs to be replaced by the distinction between cultural and technological science, both forms of applied science.

All the points that I made in connection with technological science apply with equal force – indeed with far greater force – to cultural science. *All* the human failings we discerned in technological science – as a result of the attempt to confine science within the straightjacket of irrational, unrigorous ideals for scientific rigour – can be discerned far more dramatically and deplorably within cultural science. Indeed, the situation is so deplorable, so shocking, that the very idea that cultural science should have a human point to it, should be concerned directly and actively with making a contribution to the flowering of human lives – this very idea seems to be entirely unheard of, as far as most scientists are concerned. You yourself revealed this dramatically in the objection you raised just now. "Pure" science, as you called it, is of value *in itself* you say: obviously, the idea that "pure" science could be of value because it is of value to *people in their lives*, is an idea quite impossible for you even to consider.

The whole point of science, from a cultural standpoint, is in the end to be of assistance to us, to you, me, and all the others, as people, in our attempts to improve our understanding, our appreciation, our enjoyment of our world. The whole point of cultural science, we may say, is to encourage us to have a more trusting and loving relationship with the world – a relationship that is not based on fear, on frustration, on bitterness, on desperate terror, but rather is based on calm, realistic, unblinkered, confidence and trust.

And does modern science really help, as far as most of us are concerned, in this respect? Does science really encourage us to put our trust in the world, in life? Yes, you ought to have caught the tone of my voice. These are meant to be angry and bitter questions.

The kindest thing that can perhaps be said of modern science is that as far as most people are concerned, science is in practice simply *irrelevant*. Life can be enjoyed: life can be miserable. There are enjoyable, beautiful, enticing, delightful, appealing, pleasant things and circumstances in one's surroundings; and there are unpleasant things, noisy, ugly, smelly, dirty, frightening, drab things. Technology, so overwhelmingly a part of our modern world, contributes both to the good and the bad, the pleasant and the ugly. Certainly it is important: it is present; it is *there*. But cultural science? "Pure" science? "What's that got to do with it? Isn't that just something those strange chaps, the scientists, do in their laboratories amongst their test-tubes, electronic equipment, and tangles of wire? Does "pure" science make a contribution to my enjoyment of life? That stuff about atoms and electrons and DNA and all that? What rubbish! Never give it a thought myself!"

And suppose – just suppose – we do take science seriously, as presenting to us a vision of reality, an opening up of vistas onto aspects of reality, that are rich and strange and beautiful. In practice, is our experience, our perception, of our world enriched and enhanced?

We begin, let us imagine, with the limited perspectives of a child. Our world is rich, dramatic, intense, beautiful, frightening, full of colour, strange tastes and smells, smiles, mystery. And now along comes solemn old Science to tell us how things really are. And this is what we hear!

"The universe is immense, both in space and time. Everything that occurs in the universe does so in accordance with precise (probabilistic) laws. The ultimate reality is the world as described by theoretical physics. However, on at least one cosmologically minute body encircling a star – namely Earth – various physical processes have occurred which have led to the development of self-replicating chemical systems of various diverse forms. In particular, one type of these self-replicating physio-chemical systems – *homo sapiens* – has interacted with the physical environment in various ways to produce various "artefacts" which may be termed "buildings", "ships", "aeroplanes", "cars", "books", "clothes", "furniture", and so on. All this however amounts to no more than complex arrangements of physical entities interacting in accordance with precise (probabilistic) physical laws. Various aspects of what there is may be studied by sociology, psychology, economics, zoology, botany, anatomy, neurology, geology, and so on: but in the end all these sciences need to be understood as more or less specialized branches of physics. Thus neurology studies those complex physical systems we call central nervous systems. Economics studies extremely abstract aspects of the way in which specific complex physical systems – *homo sapiens* – interact with each other. And so on.

"There is no "meaning" or "significance" to any of this. Meaning, significance or value are purely personal subjective things having no real existence "out there" in the world. Our human feelings, hopes, desires, fears, aspirations, impulses, have no significance whatsoever when it comes to the question of what actually exists in the world. If we wish to get a clear view of how things are, of what actually exists in the world, then we must throw away our merely personal, subjective feelings – themselves to be explained and understood in terms of physiology and neurology – and see things in the cold light of the intellect alone. If we do this, we will realise that human concerns and interests, human feelings and values are no more than features of certain rather unusual arrangements of rather complex molecules, somewhat

143

misleadingly described. The scientific view of the world can provide us with no reason why a special significance should be attached to these arrangements of complex molecules: the fact that we do attach such significance to them is to be explained as being no more than an expression of our bias, our egotism – for of course we *are* these arrangements of complex molecules. The human view of the world arises from no more than inflated egotism: it arises from giving a special significance to the movements, the discriminatory powers, of certain specialized physical systems. From an objective standpoint, we have no more significance than naturally occurring robots or computers – no more significance than other self-replicating arrangements of molecules."

There! Do you find that an enriching view of the world? To be told that all meaning, all value, all significance, all beauty, is purely subjective, a kind of emotional hallucination, having no objective existence out there, in the world? Is that how your world is? The world of a madman?

My God. My God. People speak of technological pollution. What of this utterly horrifying and total *cultural* pollution? You desecrate our world, in the name of your "science", your "reason" and your "objectivity". You construct the vision of a maniac – I mean it literally, the vision of a maniac, the vision of someone in emotional mental extremity, at the point of suicide – and you tell us flatly, calmly, reasonably: Yes, this is how the world is. This is what actually exists. All the rest is just your own personal, subjective, internal interpretation.

You tell me that what I have described is not really *the* scientific view of things? Not all scientists agree with it? Of course not! You are not all mad, thank God. But there it sits, quietly, this utterly desecrated vision of the world, in practice informing and influencing scientific research, in all kinds of ways, rarely explicitly articulated, precisely because it *is* a piece of metaphysics, a *vision* of things, and hence lacking in respectable scientific status, according to standard empiricism. The impact of the vision is all the more effective for being quietly presupposed rather than openly advocated. It is always the unspoken message that goes home!

144

And let me tell you something, my friend. It is rubbish. Rubbish. Lunacy. Idiocy. Here. Come here.

(At this point the Philosopher grips the Scientist by the arm and drags him to the window. The Philosopher seems to be boiling within from some extraordinary, barely contained fury. He hardly seems to be aware of what he is doing. For the first time, the Scientist really does wonder, with a twinge of anxiety, for his friend's sanity. Abruptly, the Philosopher flings open the window.)

There! That is my world. Out there. I don't know what bit of dried up, desecrated husk of a cosmos *you* live in, but that is my world. It is rich, mighty, dramatic. Look at it, for Christ's sake. Look at those great bruised clouds. Look at the air snatching at the trees. Yes. There are storms, hurricanes, volcanoes, earthquakes. There are Spring forests. There is snow. Deathly hush. Look! I mean it. Not at me. Out there! People are dying. There are people out there, lost, removed from themselves, locked in frozen wastelands of desperation. People are hungry. Mad. Terrified. Bored. Lonely. Hurt beyond description. You think I exaggerate? Oh, we get by, we get by, just about. Our world is poised for self-destruction. The mighty arsenal of nuclear fireworks, poised, ready at a moment's miscalculation, to destroy us for ever. You know how it goes? Skin turning to charcoal. The smell. Have you burnt yourself? Have you? Yes, yes, I know, I can see it: there are other things as well. There is laughter, goodness, kindness; there are such beautiful things. There are people in love. There is music. There is such abundance of generosity, warmth, courage. And we keep going, don't we?

I know. You think I am raving. I *am* raving. I am not telling you anything you don't know. But don't you see what I am saying? Your "scientific view of the world", so-called, is a lunacy, a horror, a madness. And if I have to scream to make you notice, then I will *scream*. The world is charged full of meaning, significance, colour, beauty, drama, pain, wonder, horror, death, love, happiness, despair, warmth, friendship, delight, pleasure, misery, hunger, weariness, gentleness, brutality, intelligence, sensitivity, dullness, passion, grace, madness, rage, disease, patience, courage, tenacity. All these things *exist*! Can't you get

145

that through your stupid thick skull? They exist. They are there! Out there, in the world. If your scientific measuring instruments, and your theories, don't detect and notice the existence of these things, then so much the worse for your instruments – and your theories. You are just ignoring reality. You are blind to reality. Well. Be blind. But don't go foisting it onto the rest of us.

Oh, dear; what is the point. I am sorry. I don't suppose you have any idea at all what that was all about, do you? (The Philosopher collapses wearily into a chair.)

SCIENTIST: (Cautiously) Well ….. I'm not absolutely sure that I know what you're getting at –

PHILOSOPHER: My thesis – as my colleagues would say – is really very simple. It can be put like this. The whole point of science, from a cultural standpoint, is to make itself available to us in our lives, so that we can enrich our lives, broaden and deepen our experience of the world, unroll our horizons, confront reality both more honestly, more perceptively and truthfully on the one hand, and at the same time with more trust, with greater enjoyment, even perhaps with something approaching to love, on the other hand. Modern science really can be exploited, in a personal way, to achieve these personal objectives. But at the moment it is extremely difficult to *use* science in this kind of way. All kinds of *problems* lie in the way of exploiting science in this kind of personal way, for the enrichment of one's life, the enrichment of one's vision of the world.

As science has progressed, we have discovered, we believe, various things about the world which have shocked and appalled us – especially given our previous convictions. Such things as the discovery that the earth goes round the sun; the discovery that natural phenomena appear to occur in accordance with precise mathematical laws, the world thus being *impersonal*, in no way requiring the supervision of a personal God; the discovery that we are descended from apes; the discovery that everything is made of atoms, of fundamental physical particles, including ourselves; these discoveries have all been, at various times, profoundly disturbing.

146

Even worse, there are various methodological, philosophical and metaphysical ideas, floating around just below the surface of science (like killer sharks) which have even more appalling consequences when experienced in personal terms. Worst of these is perhaps the assumption, implicit in the "scientific view of the world" that I sketched just now, that reality is such that its nature can only be determined by completely impersonal, de-emotionalized, non-sensual, non-experiential, "scientific" observation and experimentation. This assumption is an absurdity. If beautiful, horrifying, lovely, painful, amusing, ugly, pleasant, delicate, boring, courageous, stupid, noble things exist in the world, then *of course* we have to use our feelings, our emotional reactions, to discover these things, *see* them, notice them. The depersonalized, de-emotionalized "experience" of science will, of course, never come across these things just as the de-sensualised "experience" of science can never come across colours, sounds, smells, tastes, as experienced, as *seen*, by us.

SCIENTIST: But aren't qualities like "beauty" and "ugliness" inevitably subjective, in the eye of the beholder, just because people *disagree* about what is beautiful, what is ugly?

PHILOSOPHER: Not at all. Who says that we can always reach *agreement* about what objectively exists? (The old authoritarian concept of knowledge rears its head once again!) It is very easy to understand why different people come to different judgements concerning aesthetic and moral qualities (as we may call them, for a laugh). First, some people may be emotionally blind, just as some people are colour blind. We are all perhaps, at times, emotionally blind: dulled within ourselves, so that things which ordinarily call forth an emotional response from us, now do not. But second, and perhaps more important, people hold different theories about reality and about human reality in particular, and as a result seem to experience different things on the same occasion. In science, what our observations tell us depends profoundly on the theories we hold. Well, the same thing goes for life. That is why a human act which may seem noble and courageous to one person may seem despicable and cowardly to another person. What is a good deed to one may be appalling, self-seeking hypocrisy to another.

The fact that such decisive points of disagreement are all too common in life does not mean at all that the disagreements are not about what is objectively *there*, in the world. One finds precisely the same kind of radical disagreements in cosmology, for example; and yet no one wants to suggest, as a result, that the cosmos is a subjective phenomenon.

Back to the point I was making earlier. All kinds of problems, difficulties and traps lie in wait for anyone who wishes to exploit science in order to extend and enrich his personal experience of the world. But these problems can be solved, these difficulties overcome, these traps sprung. At least I, *personally*, have found this to be possible. Ruthlessly, greedily, mercilessly, I personally exploit all the labour of scientists to explore and enrich my world. After all kinds of difficulties and struggles, I have discovered how to take just what I need, what I want and value, separating out the precious essence from the irrelevancy. From the intricate mathematical formulae of modern physics I have learned how, to some extent, to pluck out the *ideas*. From the jargon of biology, and amongst the unvoiced, foolish assumptions (as they seem to me) I plunder what is for me precious and valuable. It is quite clear to me from the way the whole thing is set up that no one is *meant* to treat science in this cavalier selfish fashion. I rarely find other academics engaged in the same game: they stick to their specialized subjects; they would regard what I do as a combination of insufferable arrogance and academic suicide. Certainly none of my colleagues – the philosophers, the philosophers of science – are engaged in this marvellous, delightful, but at times rather gruelling pastime: they all dutifully stick to their specialized interest. There is, however, one exception: Karl Popper. It is at times a rather lonely pastime; and at times I am rather buffeted by specialists who resent my keen, critical, wholly unprofessional interest in what they are doing.

And gradually it has begun to dawn on me: what I do is the real thing. This is what science is *for*, from a cultural standpoint. The really important, valuable, worthwhile enterprise is what might be called natural philosophy, and human philosophy: the personal exploration of the world, natural and human, and the sharing of one's ideas, one's discoveries, one's experiences, one's problems and difficulties, with one's friends. What could be more wonderful than this? There are other things in life, of course. But it is this, for me, which, to a considerable extent, makes life worth living.

And my discovery fills me with dismay, with something approaching horror. For almost no one seems to be treating natural philosophy and human philosophy in the kind of way that I described, as something like a first intellectual priority, close to the very heart of one's life. It does not seem to be a recognized activity. My calm assumption that science is to be *used* in this way so often seems to upset scientific and academic experts, in that they seem to find the assumption to be both naive and arrogant. Everything in their training, their professional etiquette, runs counter to what I do.

And what are the consequences of these almost universally held scientific and academic attitudes? It is almost impossible for anyone to use science, from a cultural standpoint. Scientists once again are so busy accumulating new technical knowledge that they have forgotten to ask: *Why* are we accumulating all this knowledge? What is the human point of it all? Who can use it, enjoy it, value it? It is all very well for me. I am more than willing to throw away a successful academic career in order to get down to the real thing. I am aware that very few people anywhere are doing what I am doing. I am exploring virgin territory. It is quite extraordinarily exciting. New problems, new vistas, new riches open up before my amazed eyes. (You can call all these delusions if you want: I don't mind.) I once made a list of the problems that interest me, for my students. In a couple of hours, I had about sixty problems. It saddens me, of course, that there are so few people to *talk* to about all this, to share all this with. It is very difficult, boring and time consuming writing up my discoveries, my ideas for publication in philosophy journals, just because what my colleagues assume philosophy to be is so different from what I assume it to be. So mostly I do not make the attempt.

One major consideration casts a shadow over all this: the state of the world. The misery in the world. Along with many others, I feel an obligation to try to be of help. I believe in philosophical technology just as much as I believe in pure philosophy, as you would call it.

But the point I want to make is this. I am very lucky. I have plenty of time and opportunity to follow up my interests. If I become interested in the nature of mathematics, for example, off I can go to ransack the library for any relevant articles or books that I can find. No one tells me that I cannot possibly follow up such an interest because I am not qualified, I do not know enough mathematics, I do not have a first degree in mathematics. If I was a Ph.D research student, however, I would not be allowed for one moment to pursue this interest. For undergraduates the thing is hopeless. For people outside the university set-up the thing is very difficult, very discouraging. The mathematicians are too busy doing their mathematics to worry their heads about such nonsense. And so it is left to me!

But if cultural science, "pure" science, as you call it, were truly rigorous, truly objective, then it would be a commonplace idea that the whole point of cultural science is to be of assistance to people in their exploration of their world, in the kind of way that I have described. This would be something easily available and understandable, not just to the odd academic freak, tucked away in a philosophy department, but to *anyone*, whatever their knowledge, their background, their "I.Q.". Scientific, intellectual problems would be clearly understood as personal problems, problems arising in one's life, openly available to people, and not the exclusive property of experts. Scientific knowledge and understanding would be understood to have a personal, social character. One would not speak vaguely of increasing "man's" knowledge, "man's" understanding. The concern would be with the knowledge, understanding and appreciation of *people*.

So: do you see now why I regard the cultural case to be essentially the same as the technological case, apart from being, if anything, worse? Scientists seek knowledge, but do not ask the elementary question: Why? What is the human point of all this knowledge we are accumulating? How does science from a cultural standpoint, serve to help enrich human life? Precisely as a result of conceiving of "pure" research as in itself seeking to help promote human welfare, be of value to people, all the problems and barriers between science and people could gradually be overcome (arrow C of diagram 4.)

SCIENTIST: I would have thought that many scientists have something of the same attitude to science that you seem to have – a genuine *personal* interest, a part of a personal curiosity about the world.

PHILOSOPHER: I agree. And that very point is an indication of just how selfish, how decadent, how subjective science has become, when viewed as a cultural phenomenon.

Suppose scientists developed all kinds of technological devices – medicines, cars, electric lighting, television, etc., etc. – and made sure that all this was kept for the exclusive use of scientists, the rest of the population being obliged to live in caves. Well, I think we would all agree that some element of injustice is to be detected in such an arrangement. *But precisely this is the situation on the cultural level!* The only people who are supposed to be capable of being authentically preoccupied with scientific, intellectual problems are – scientists! Qualified scientists. And I think you would agree with me that the real excitement of science comes from tussling with *problems*, grappling with the *unknown*, not simply from absorbing the *known*, from being told by the appropriate expert what to think, what to believe, what to accept.

SCIENTIST: Yes.

PHILOSOPHER: It is the spirit of optimistic, delighted, awed, sceptical enquiry that surely represents science at its best. And it is just this spirit that scientists keep to themselves.

SCIENTIST: Your sentiments, your intentions and hopes are, I suppose, admirable. It's just that they seem to me to be completely unrealistic. Your idea is that it ought to be possible for more or less *anyone* to exploit the whole range of scientific knowledge in order to build one's own personal philosophy of life, one's own personal philosophy of the nature of reality. But this seems to me to be today something that is quite impossible, even for a person of quite exceptional genius, let alone for an ordinary mortal like you or me. Once upon a time, when scientific knowledge was far less

151

extensive and technical, a few rare men of genius, such as Leonardo, Galileo, Kepler, Newton, Descartes, Leibniz, Spinoza, could perhaps exploit science in this kind of way. Much of their work and writings can perhaps be interpreted as the outcome of such a personal exploitation of scientific knowledge. But even in those days of limited knowledge, only rare men of outstanding intellect who devoted a lifetime to the task, were capable of such an achievement. Nowadays this Renaissance ideal of the person of balanced, rounded culture in depth has disappeared for ever, even for the rare individual of genius with a lifetime available for the task. Modern scientific knowledge has become, quite simply, too vast, too technical, too intricate, for it to be conceivable that any one individual could encompass the whole domain. It has, in fact, become more or less impossible for any one individual to encompass a single discipline, such as mathematics, or theoretical physics, let alone the entire corpus of scientific knowledge. The idea that an ordinary run-of-the-mill taxi-driver, novelist, postman or lawyer could achieve such a thing, without specialized training, with an ordinary run-of-the-mill intelligence, and while devoting much of his time to other pursuits, seems to me to be, speaking frankly, the height of absurdity.

PHILOSOPHER: There you are, you see. A serious problem arises in connection with bringing science to people, to life, and you are ready to abandon the whole endeavour as hopeless. Clearly, the immense success of modern science creates serious problems when it comes to rendering scientific knowledge fit for human consumption. Our response should be: right, let's try and get clear what the problem is, so that we can *solve* it. Instead people adopt your attitude. The situation is hopeless; nothing can be done. And as a result, no thought, no care, no attention is given to looking for solutions to these problems. Scientists just go on developing more and more technical knowledge mindlessly, without any idea as to why they are doing what they are doing, making the situation progressively worse and worse. One might almost say: the immense success of modern science has almost brought about the death of science, when science is conceived of as a cultural enterprise, experienced, shared and used by people.

The truth of the matter might be put like this. If our modern scientific knowledge is recorded and embodied in such a way that it is utterly inconceivable that any one individual, even an individual of genius, can any more oversee the whole enterprise, and grasp what it is all about, then we have quite simply lost control of this human product. We human beings are no longer in charge: we no longer use science; science uses us.

The idea of scientific technology taking us over, and using us, instead of our using it, as a means to the realization of *our* desirable human ends, is in these days a relatively familiar nightmare, expressed in a number of works of science fiction, and other writings. (An early and especially beautiful expression of this idea is to be found in Butler's *Erewhon*.[14]) The possibility that something like this has already happened as far as scientific *culture* is concerned, scarcely seems ever to be noticed. Or rather because the takeover of scientific culture is usually deemed to be absolutely inevitable, the price that we are bound to pay for scientific advance, the fact of the takeover is scarcely even lamented.

If scientific knowledge has grown to grotesquely unwieldy and unusable proportions, then it has become a matter of the first priority, a matter of the most absolute urgency, that we seek to put the thing right, by redesigning and rearranging the whole way in which our scientific knowledge is represented, so that it does become fit for human use. In the present situation, the more we add to the technicality, intricacy and specialization of scientific knowledge, then the more we sabotage real scientific progress, when that progress is assessed in human terms. At present conventional scientific progress serves only further to *undermine* real, human, scientific progress.

What we need to do, quite clearly, is to press and squeeze the valuable essence of scientific knowledge and understanding from the vast intricacy of its technical expression, so that it becomes humanly comprehensible and usable. The essence of scientific knowledge and understanding, the fundamental ideas, discoveries, problems, aims, speculations, hopes, aspirations, theories and insights need to be rendered up in simple, clear, delightful, entrancing prose, available for the enjoyment of anyone. That which may be needed for the building of one's own philosophy of life, one's own philosophy of reality, should be available and accessible to all.

A technical science that has been intelligently and sensitively designed for *people* might be conceived of in the following terms.

The public record of scientific enquiry, its achievements and failures, its problems, aims, methods and aspirations is, let us imagine, arranged in a series of concentric circles. At the centre, representing the heart of science, the fundamental achievements, problems and aspirations, there is, we may imagine, a little book which a child of eight can understand and take delight in. It would be of decisive importance that this little book records not only the essence of what we know, but also the essence of what we conjecture we do not know, but divine and hope that we can discover. The book would lay as much emphasis on fundamental unsolved problems, on our best conjectures about the domain of our ignorance, on our best ideas for aims for future research, as on what has already been achieved. This little book is, as it were, at the centre of scientific concern; it embodies the fundamentals, the essence of scientific knowledge. The highest scientific honour goes to the authors of this little book; the fiercest scientific controversies centre on what should go into the little book. The writing of such a book would require a profound mastery of science in all its aspects, a deep sense of historical perspective, a sure grasp of fundamentals, an awareness of the most basic and lasting achievements, problems and concerns of science, as opposed to that which is merely ephemeral or fashionable. The evolution of this little book over the years would provide the true record of progress in scientific knowledge and understanding.

Radiating out from this central little book in all directions, there would be increasingly more technical, specialized treatments of various aspects and branches of science, arranged in such a way that the uninitiated, having read only the central little book, would be able to find his way, picking and choosing that which corresponds to *his* interests, problems and concerns. Scientific knowledge and understanding would be arranged for the convenience of people, so that they can find their own way around with profit and enjoyment, as opposed to science being arranged solely for the convenience of specialists and experts. On the outer rim of recorded scientific knowledge, in specialized journals, new results, developments, theories would be recorded: but nothing new of fundamental importance would be properly "published" in the literature before it had found its way to its proper representation in the little book.

At present, scientific knowledge is arranged in a way that is very different from this humanistic ideal. A non-scientist, eager to *use* science in order to extend and enhance his experience, his vision, his knowledge and understanding of the world around him, will quickly find himself lost in an ocean of abstruse jargon and technicalities. Far from returning us to the world around us, with renewed interest and understanding, science, in practice, as it is at present arranged, seems to take us further and further away from the world, from reality, into a maze of inexplicable abstractions. How can one hold up a modern theory of theoretical physics to see what the world looks like through it, if the theory is itself so opaque to the understanding that one cannot even see *into* the theory, let alone *through* the theory to the world beyond? Scientific knowledge is contained in the form of an ever-increasing number of specialized disciplines, boxes of knowledge each of which might take a lifetime to absorb. How, then, can one conceivably get a view of the inter-connections, the overall patterns and themes? The *world* does not present itself to us as neatly divided up into theoretical physics, astronomy, cosmology, astrophysics, zoology, botany, physiology, neurology, molecular biology, inorganic chemistry, etc., etc. It is absurd to try to see the world through the multi-faceted insect eyes of all these disparate disciplines when the very way in which these disciplines are sub-divided and distinguished often only has academic, administrative significance.

At present, we seem to be engaged in creating a vast mansion of knowledge which is rapidly becoming ever more intricate and convoluted, ever more removed from and indifferent to the needs and interests of people, it being entirely forgotten who is supposed to live in this mansion. It is as if we build our science for some

155

mythical super-intelligence of the distant future, whose only passion is to *know*, and who alone will be able to master all the multitude of specialities and technicalities of modern and future science, in order to survey and comprehend the whole cathedral. We need to rearrange things so that ordinary people – you and I – can today comprehend this vast edifice, at least as far as essentials are concerned, and so far as our own interests take us. We need a science that is fit for human habitation. At present, we seem to be engaged in building a cultural monstrosity that has perhaps as much relevance to our needs as the Pyramids had to the needs of the average ancient Egyptian.

When judged from a cultural standpoint, science is, it seems, for *specialists*, not for you and me. And this indicates just how decadent modern science has become, when viewed as a part of our general culture. A theatre only understood and appreciated by theatricals, a poetry only understood and enjoyed by professional poets, a literature only understood by professional novelists, would immediately be condemned as appallingly decadent and narcissistic. And yet, when it comes to science, this kind of cultural decadence and narcissism is entirely tolerated, and even seems scarcely to be noticed, let alone criticized and condemned. Indeed, scientists are inclined to lay the blame on non-scientists, rather than on the way in which professional scientists themselves exhibit and share their ideas and discoveries. It does not seem to occur to scientists that an essential *professional* obligation is to render up that which is of greatest value in science in a form that is comprehensible, interesting and exciting, and in a way that stimulates curiosity and is not off-puttingly patronizing. The whole attitude of specialist scientists to their enterprise seems to be quite different from the attitude of artists to their art. Novelists, poets, musicians, painters, do not usually act as if they *owned* their art: art is public property, available for public enjoyment and criticism. Scientists, on the other hand, seem to regard science as their exclusive possession: outside criticism is ignored or resented. The centre of gravity of science does not exist in the community as a whole: it lies within the exclusive circle of specialist scientists.

The great danger of developing science in this exclusively specialized kind of way is that science itself must in the end begin to suffer and decay, even when judged from a relatively conventional *scientific* standpoint, leaving aside the kind of humanistic, cultural standpoint that we are considering here. Just because the world itself is not neatly divided up in a way which corresponds to the distinctions and categories of specialized academic sciences, the great, fundamental scientific problems of knowledge and understanding do not fit neatly into these pre-arranged academic boxes. Such problems often cut across conventional academic disciplines: they can only be formulated and understood by people who are prepared to step back from obsessively specialized science – precisely in order to get a glimpse of broader perspectives. Science will only tackle, and grope towards, solutions to these broader, fundamental problems of knowledge and understanding if there are scientists with specialized knowledge and skills who are capable of freeing themselves from the highly restricted vision of the specialist simply in order to notice the existence of the broader problems. But this becomes something that it is increasingly difficult to do as science is pursued, recorded and taught in an exclusively specialist fashion. If professional scientific propriety demands that one stays in one's specialist box, and does not wander into disciplines in which one has received no specialized training, then it becomes almost impossible for research scientists to get a glimpse of the broad, fundamental problems. Knowledge and understanding itself becomes increasingly fragmented and disconnected. The very *idea* that science ought to seek to solve fundamental problems of understanding begins to disappear. Retaining an interest in such problems undermines one's scientific standing and status, as an expert. The fundamental scientific objective of seeking to improve our knowledge and understanding of the world around us degenerates into the task of solving specialized puzzles within the specialized disciplines.

There is thus an urgent need to put into practice something like the "little book" idea for the representation of scientific knowledge for the sake of the health, rigour and progress of science itself. Specialist scientists need to retain an appreciation of the fundamental objectives, aspirations and problems of scientific enquiry, formulated in simple, non-technical terms, simply in order to retain and develop, an understanding of how their specialized problems and interests take their place within broader, more fundamental scientific concerns. Specialists may well have much to learn from interested non-specialists, and from their students, just because an outcome of specialization may be that one loses sight of the wood for the trees – or even for the twigs! It is not always easy to remember – just because it can be a disconcerting, humbling experience – that one's specialized interests do represent twigs.

Considerations somewhat similar to these have moved Popper to remark recently: "If the many, the specialists, gain the day, it will be the end of science as we know It – of great science. It will be a spiritual catastrophe comparable in its consequences to nuclear armament".[15]

SCIENTIST: You philosophers seem to be all the same. Full of hysterical outbursts.

PHILOSOPHER: Yes, perhaps we do.

SCIENTIST: In any case, irrespective of the question of whether I *agree* with you about all this, I find your basic idea really quite appealing – your idea, that is, that a more truly *rigorous* science would also be a science more directly and consciously concerned to be of use and of value to people.

PHILOSOPHER: I am delighted.

SCIENTIST: And now I must go.

PHILOSOPHER: One last small point.

SCIENTIST: No really, I —

PHILOSOPHER: It won't take a moment. I merely wish to say that the *cultural* human failings of modern science are, in my view, far more fundamental and serious in the long run than the *technological* human failings. Ideally, science should be there to help us improve our understanding and appreciation of the world around us. At the moment, for most people, science achieves, if anything, the exact opposite of this. It is almost as if vast, esoteric, authoritarian, technical, unintelligible Science interposes itself as an impenetrable barrier between person and world. One entirely understandable response to this situation is to say: science is not really about the world, about *reality*. It is just a game played by scientists. Even scientists themselves have a tendency to opt for this false solution, in that they concern themselves with "models", with algorithms for predicting phenomena, with little-understood mathematical formulae, instead of concerning themselves with reality itself, theories, ideas, being our speculative gestures towards reality, our adult equivalent of a child's "Look! The moon!".

It is not true. Science is concerned with reality. And as long as science can only be understood by a whole bevy of experts – who often cannot understand each other – then the consequence is that none of us can any more understand our world. Only the *experts*, all together, understand the world (ha!). No individual person knows any more how to put the pieces together.

And it is surely quite clear that the consequences of this must be devastating. It is the end of any hope for democracy. If we do not understand our world any more, but have to take on trust what the experts tell us about our world, how can we possibly be in a position to make intelligent, sensible, political, social and moral decisions for ourselves? We have ceased to be free, responsible people; we have become manipulatable zombies. Our *technological* troubles, I would say, are but a product, an offspring, of our *cultural* troubles, associated with science.

SCIENTIST: Right! Good! That's it. I'm off.

(Outside the window, two workmen are mending a drainpipe in the fine drizzle that is adrift in the air.)

OLDER MAN: (Indicating with his thumb the open window from which the sound of voices has been rolling out) Here, they don't half go on and on, these university types.

YOUNGER MAN: Yeah.

OLDER MAN: Not a bad job though – being paid for nattering away about whatever takes your fancy.

YOUNGER MAN: No, man, you get your head really screwed up playing those kind of academic games. You forget where it's all at.

OLDER MAN: Oh! (He wonders what "where it's all at" is supposed to mean.) All the same. A cushy job, being paid for doing what you want to do anyway.

YOUNGER MAN: That's the way the system works, man. You enjoy your job? Then well pay you well. You're doing something that's as tedious as hell? Then we'll give you a pittance.

OLDER MAN: (Grunts) Tea?

(They wander off for their tea break.)

CHAPTER EIGHT

REASON AND PHILOSOPHIES OF LIFE

(Some days later)

PHILOSOPHER: Now comes the really big step!

SCIENTIST: Oh?

PHILOSOPHER: Up till now we have just been clearing up a few preliminary matters. What I have to propose now is the crucial step towards recognizing that person-centred science, Pygmy science, represents a truly *rigorous* ideal for science.

SCIENTIST: I see!

PHILOSOPHER: What I have to propose is a completely *new* way of applying science to life, to our multifarious personal, social problems and pursuits. My claim is that this will enable us to get into *life* something of the progressively successful quality that is found so strikingly in *science*, on the intellectual level.

SCIENTIST: (Grinning: not believing a word of it) Ah, ha!

PHILOSOPHER: What I have to propose is in essence a new ideal for *reason*, a more *rigorous* ideal for reason, which is at one and the same time an ideal sensitively adapted to helping *us* to discover and to attain that which we really want to attain. At the same time, this provides a completely new idea for the aims and methods of the social sciences. At last, it becomes possible to put the social sciences on a truly rigorous, rational, scientific footing: and, as we shall see, doing this will transform the social sciences into something of immense help and value for our various personal and social problems and pursuits. In a sense the social sciences will turn out to be more *fundamental* than the biological and natural sciences – a complete reversal of how they are usually viewed. The biological and natural sciences become, almost, specialized offshoots of the central, fundamental social sciences.

161

SCIENTIST: (With mock gravity) I see. Is that all?

PHILOSOPHER: By no means! Our new aim-oriented conception of rationality – of universal application, to all that we do – enables us to develop a striking new interpretation and generalization of some key ideas of Freud and psycho-analytic theory. These ideas are interpreted as rationalistic, methodological ideas, of universal applicability to all aim pursuing activities, and not just to people, to the human psyche. This reinterpretation utterly transforms the standing of Freud's ideas. Instead of being obliged to hold that psychoanalytic theory does not really come up to the exacting standards demanded by science, we can turn the tables utterly: we can argue that it is *science* which does not come up to the exacting standards of rationalistic psychoanalytic theory. For it will turn out that, at present, science suffers from what may be called rationalistic neurosis, a purely rationalistic, methodological condition which can beset any aim pursuing enterprise that misrepresents to itself its basic aims. All our previous criticisms of science – its lack of rigour, its human failings – in the end can be traced back to this simple cause: science suffers from rationalistic neurosis.

SCIENTIST: Back to that again! You still want to put science on the psychoanalyst's couch – only now it is a *methodological* psychoanalyst's couch?

PHILOSOPHER: Exactly!

SCIENTIST: I see. But come. That can't be all. There must be more to it than this.

PHILOSOPHER: Yes, there is. Aim-oriented rationalism leads to the conclusion that at the centre of our intellectual attention we should put – philosophies of life, ideas concerning the aims and methods of life, just that which traditionally has been so despised, from an intellectual, rationalistic standpoint, as incorporating a confused mixture of ideas concerning facts and values. But even more, the centre of our intellectual attention should be *life*; the aim of scientific, intellectual enquiry should be to help promote the flowering of life, consideration of *philosophies* of life being but an offshoot of this, a means to that end.

SCIENTIST: Go on.

PHILOSOPHER: Aim-oriented rationalism provides a methodology for rationally *appraising* philosophies of life against our own personal experience of life. We can, as it were, test philosophies of life against our own personal experience, and use our personal experience to develop our own philosophy of life that suits us the best – in much the same way in which scientists test and develop scientific theories. Indeed, we may even hold that, from a truly rational, rigorous standpoint, the testing and developing of *scientific* theories is but an offshoot, an aspect, of the personal, cooperative testing and developing of philosophies of life.

SCIENTIST: I see. (He doesn't, of course, at all) And what else is there?

PHILOSOPHER: Well! Let me see. This new theory of rationality – aim-oriented rationalism – is not only more rigorous than previous theories; it is also, at one and the same time, a theory of *rational* creativity. Creativity, discovery, instead of being something essentially irrational, ineffable, (as it tends to be with respect to existing conceptions of reason) turns out to be a rational process, capable of being genuinely helped by reason.

SCIENTIST: (Ostentatiously and wickedly suppressing a yawn) Anything else?

PHILOSOPHER: Yes. Aim-oriented rationalism is also a general theory, or methodology of problem solving, and at the same time a theory of learning, with important implications for education, which puts action, play, before thought.

SCIENTIST: But this can't be all, *surely?*

PHILOSOPHER: No. Let me see. Oh yes! Aim-oriented rationalism provides us with a general theory for the rational, desirable use of culture, science, art, tradition, institutions, customs, ideals and values, so that we may arrange these things helpfully and fruitfully around our lives, our actions, to encourage our lives to develop in fruitful, desirable, fulfilling directions.

SCIENTIST: But this, surely, is implicit in what you have already said?

163

PHILOSOPHER: Yes, I suppose it is in a way. Well, then. Aim-oriented rationalism enables us to inter-relate, in both a rational and a desirable way, thought and feeling, idea and desire, mind and heart, the objective and the subjective, science and art, intellect and imagination, the impersonal and the personal –

SCIENTIST: Come, come, come. This again is surely implicit in what you have already said. You seem to be scraping the barrel a bit. Isn't there anything else?

PHILOSOPHER: (Becoming now a bit desperate) Aim-oriented rationalism can help us to bring about a harmonious relationship between ourselves and Nature.

SCIENTIST: There must be more to it than that!

PHILOSOPHER: (Now really scraping the barrel) It is the answer to all our troubles.

SCIENTIST: Oh, *that*!

PHILOSOPHER: (With a triumphant laugh) *It's all common sense!*

SCIENTIST: I am glad to hear it. But, I must admit, I am disappointed. I was, I admit, hoping for something really BIG: for example, something like a grand synthesis of, let us say, Buddhism, Christianity, Rationalism, Romanticism, Physics, Biology, Liberalism and Marxism.

PHILOSOPHER: (Grinning) Well, yes, there is that too.

SCIENTIST: O.K. then. I'll lend you half an ear. But please don't let your exposition drag on for too long: I don't want to miss my lunch.

PHILOSOPHER: Right! Off we go, then.

So far, we have been discussing ways in which science can be of use, of value, for *life*. I have argued that, at present, science is not very satisfactorily being of service to us in our lives because science has been trapped within seriously unrigorous conceptions of scientific rigour. Release science from the straightjacket of standard empiricism, and not only will science itself flourish, in terms of relatively conventional *intellectual* criteria of scientific progress, but in addition, and surely far more important, the relevance, the value, the use of science for our *lives* will flourish as well. These are indeed but two sides of the same coin. We will have a truly rigorous science, which is also sensitively, delightfully and compassionately responsive to human problems, needs and aspirations, practically efficacious in helping us –

SCIENTIST: I was not asleep, you know, during those earlier discussions.

PHILOSOPHER: Yes, I am sorry. The point I was trying to lead up to is this. So far we have discussed essentially just *two* ways in which science can be of value to us in our lives. On the one hand, science can be of value via *technology*: technological developments, made possible by scientific knowledge, can help us to solve our life problems, realise desirable, personal, social objectives. And on the other hand, science *could* be of value culturally; the discoveries, the ideas, the theories, the explorations of science can, in principle, be used by us personally, in our lives, to extend and deepen our personal experience, knowledge, understanding and appreciation of the world around us – thus enriching our lives.

SCIENTIST: (Impatiently) Yes, yes, yes.

PHILOSOPHER: (Unperturbed) Both uses of science involve essentially using the *products* of scientific research.

Now I am going to continue the whole argument: but I am going to consider a quite different way in which we may seek to apply science to the task of enriching our lives.

My suggestion? That we should apply not the products of scientific enquiry, but rather that we should apply the *methods* of scientific enquiry to all our various personal, social, institutional activities, pursuits and enterprises.

We can look at it this way. One of the most striking features of science is precisely its highly successful progressive character, within its own field. As long as we conceive of the intellectual progress of science in relatively conventional terms (hovering somewhere between standard empiricism and aim-oriented empiricism) and as long as we restrict our attention to the natural and the biological sciences, ignoring for the moment the social sciences (which seem to have met with much less generally acknowledged success), then there can be no doubt that scientific enquiry has met with quite staggering success, in its own field. And this success seems to be essentially progressive and accumulative in character. Quite fundamental revolutions can occur: theories, once accepted as true, are subsequently recognized to be strictly false: but such strictly false theories – such as Newtonian mechanics – are still held to constitute essential contributions to scientific knowledge, essential stepping stones to subsequent developments. New revolutionary theories build progressively on the very theories that they replace. There is, in this way, progressive development. And this kind of progressive development seems, if anything, to become ever more rapid. As our knowledge advances, our very knowledge about how to improve our knowledge seems to advance as well.

All this seems to be in striking contrast to the rest of human life. There can be no doubt, of course, that the general quality of human life has in many ways made enormous strides forward in some parts of the world at least, during the last fifty, one hundred, or two hundred years, as a result of industrial development, the human use of scientific technology, political, institutional, educational and even moral developments. In many parts of the world, however, quite a different story needs to be told. Industrial, technological development has brought with it appalling new problems and dangers – problems of pollution, depletion of natural resources, modern armaments, the Bomb. Despite all our progress, it is not clear that the quality of our modern civilization is in all important respects an improvement over aspects of life in ancient Athens, for example, during the time of Pericles. Even the Pygmies of the rain forests of central Africa seem to have highly valuable aspects to their life which we seem somehow to have largely lost, despite all our progress and advancement. It is not clear how many individual people in our modern culture and society find progressively increasing richness, fulfilment and wisdom with the passing of the years. Modern music does not surmount the achievements of Bach, Beethoven and Mozart; modern drama is not clearly an advance on

166

Shakespeare. We are beset by basic political, economic, social, institutional problems which, in certain respects, we seem to be no more capable of tackling successfully today than people and societies of earlier times. Nearly all of us individually, in one way or another, experience the modern world as developing in directions which seem to be frighteningly out of human, or humane, direction and control. The amazingly successful, progressive character of science scarcely seems to be a feature of the rest of human life.

And now comes my suggestion: perhaps if we could apply the *methods* of scientific enquiry, used with such amazing success within the restricted domain of science, to our broader human, social problems, we might be able to put into *life* something of the progressively successful quality that is found so strikingly within science. As a result of applying the methods of science to our pursuits in the fields of politics, industry, education, the arts, commerce, international relations, the law, our own personal lives, we might be able to get into all these diverse personal, social, institutional pursuits and enterprises some of the progressive success, when judged in human terms, that is found on the intellectual level within science.

This is the general background to the suggestion I wish to make. The *specific* suggestion I wish to make can now be put like this.

As long as we stick to conventional conceptions of scientific method, we shall not be able to extract from science a method that can be applied with very fruitful results to all our other personal, social, institutional activities, pursuits and enterprises. If we try to extract from *standard empiricism* a methodology of general applicability and value in life, we shall not come up with anything very helpful.

The reason for this, however, is that standard empiricism is a seriously defective, unrigorous methodology for science, which singularly fails to capture the implicit methodology in science which is responsible for the successful, progressive character of science. Success in science, as we have seen, has been achieved *despite* scientists' allegiance to standard empiricism, not because of

167

it. No wonder the attempt to apply standard empiricist methods to other human activities and pursuits outside science does not achieve very much that is of real human value.

The whole situation is, however, changed dramatically and profoundly if we adopt, as our conception of scientific method, not standard empiricism, but the more rigorous ideal of *aim-oriented empiricism*. This conception of scientific method, this ideal of scientific rigour, which so strikingly captures the amazing progressive character of scientific success, turns out to be of profound relevance and value for all that we do in our lives. Aim-oriented empiricism can, in short, be generalized to form a methodology, a conception of rationality, of universal application to life – a conception of reason which may be called aim-oriented rationalism. Quite suddenly it becomes humanly *desirable* in the extreme to pursue our various personal, social activities and enterprises, taking aim-oriented rationalism into account. Previous suspicions that we may have had concerning the desirability, relevance or fruitfulness of reason – for activities such as poetry, for example, or love – melt away into the air like morning mist. Quite suddenly it becomes the height of human desirability to live a rational life, develop a rational society – things which previously might well have struck horror in our hearts, and with justification.

SCIENTIST: So! You are advocating that we should seek to develop what one might call the *scientific* society, the *rational* society: and you are even claiming that it is desirable to *live* a scientific, rational, life.

PHILOSOPHER: That's it.

SCIENTIST: (In an attempt to highlight the full absurdity of the thing) You are suggesting, in other words, that we should love rationally – or even: that we should love each other *scientifically*.

PHILOSOPHER: (Unperturbed) Yes.

SCIENTIST: Well, well, well. The only thing is, I'm not sure that I *want* science, and rationality, intruding into my married life. Love, it seems to me, rests on feeling, on intuition, on instinct: it is *personal*. It has nothing to do with reason. I am sorry: rational love, *scientific* love, sounds to me like an absolute and total

168

contradiction in terms. If one attempted such a thing it could only result in a complete perversion of genuine love. The nearer to love love becomes, the less "scientific", the less "rational", it *must* be. And the more "scientific", the more "rational", love becomes, so the colder, the more impersonal, the more heartless it must be; in other words, the more *unloving* it must be.

PHILOSOPHER: But all these immediate reactions of yours only arise because you are thinking of science, of reason, in conventional, *irrational* terms. The moment we have available a more truly *rational, rigorous* ideal for reason, everything changes dramatically. Precisely the kind of objections that you have voiced to my suggestions vanish into the air. It becomes the height of desirability to live rationally. Reason becomes the servant of desire. Reason is desirable simply because it helps us to discover, and to attain, the truly desirable. It suddenly becomes positively *unloving* not to love rationally. Love *needs* reason, in order to perfect itself, in order to realise its own loving purposes.

SCIENTIST: I am sorry. I can't make head or tail of what you are talking about.

PHILOSOPHER: Well. Consider your immediate reactions to my suggestion. You said you were not sure you wanted rationality intruding into your married life. To me that remark indicates that you are still thinking of reason in authoritarian terms, as something that, ideally, decides things for us. My whole point is that such an authoritarian conception of reason is profoundly irrational, and of course profoundly undesirable. The proper, desirable, rational task for reason is to help *us* to make the decisions, the choices, that we ourselves really do want to make. The rational task for reason is to *enhance* our capacity to choose wisely and well, to enhance our capacity to choose as we really want to choose. Thus there can be no question of rationality intruding into your married life. There is only the question of your enhancing your capacities to make the choices that you really do want to make: it must of course be entirely your free decision whether you wish to exploit rationality for this purpose.

Again, you say that love can have nothing to do with reason, because love is based on feeling, and is personal. But only irrational conceptions of reason dissociate reason from feeling. As we have seen, the fundamental *intellectual* problem concerning the rationality of science – namely the problem of induction, bequeathed to us so dramatically by Hume's simple, sceptical arguments – arises precisely because science, thought, is dissociated from feeling. Absurd, unacknowledged feelings – the absurd desire to obtain power over Nature, arising from fear of Nature – create the intellectual, rationalistic problem. And in order to resolve the problem we actually need to rediscover the unacknowledged feelings underlying science; and we need to put these feelings right. Science only begins to make rational sense when we begin to realise that science is an expression of something like our trust in Nature – indeed, is something approaching an expression of our *love* for Nature, for our world. Pygmy science is *rational* science.

So you see: I could hardly be advocating that reason is somehow the enemy of feeling, or requires an absence of feeling in order to operate properly. It is all the other way round: we cannot be acting rationally if we are out of touch with our feelings. Reason, one might almost say, is feelings working themselves out, intelligently, effectively and practically, to a good conclusion. Reason counts for very little unless it is our *heart* that is rational, instinctively and impulsively.

SCIENTIST: All I can say is this. I find it altogether unbelievable that a concept of reason, with the kind of features you have just described, could be *demonstrably* more rigorous than conventional conceptions of reason and rigour. I would like to see such a demonstration! (He says this in a way which indicates clearly that he is convinced that such a thing is quite impossible.)

PHILOSOPHER: Very well. You *shall* see such a demonstration!

SCIENTIST: Before you begin, there's one preliminary point that I would like to make. It is all very well planning some perfectly rational world, a rational Utopia, in which everyone lives in accordance with the light of Reason. But how are you going to persuade people to take up your suggestion seriously? Is not irrationality an all too prevalent and potent a force in the world in practice? And are there not all kinds of people who are actively *hostile* to reason – who regard reason as almost something of a menace, something which they feel poses a threat? It seems to me that, in practice, your suggestion, your proposal, does not really amount to very much. For centuries, after all, people have been putting in a plea for reason without very much effect, as far as one can see.

PHILOSOPHER: First, I certainly don't want to persuade anyone to "live by the light of reason". That would do violence to my whole intention, which is merely: to open up a *possibility*, which may not have been clearly noticed before – a possibility which people may perhaps, of their own volition, wish to take up, to explore. My concern is to open up a new possibility, thus enhancing the range of choices that lie before us, in this way increasing individual freedom: I certainly don't want to close down any possibilities. What I *do* believe, however, is this. The possibility that I am gesturing towards here is not only a relatively new possibility, one that has not been clearly seen before: in addition, it is, I believe, an immensely *attractive* possibility, having I believe an almost universal appeal. And, in addition, it will, I believe, be especially attractive to precisely those people who have traditionally been most hostile towards "reason".

SCIENTIST: What do you mean? (The scientist has to raise his voice somewhat at this point to be heard, because of distracting scuffling noises going on outside the door.)

PHILOSOPHER: Those who have traditionally been most hostile towards reason have been those who have been most sensitive to the very serious rationalistic *defects* of traditional conceptions of reason. Unfortunately, instead of criticizing so-called rationalists for upholding irrational conceptions of reason, then going on to develop more rigorous, more rational conceptions – instead of this, tragically, they have denounced "reason" *in toto*. It has not occurred to them that what is objectionable about "rationalism" so-called is precisely its irrationality.

171

The desire oriented concept of reason that I wish to put forward will, for example, I suspect, be held by those sympathetic to the romantic movement, to capture the deepest and finest insights of romanticism – even though traditionally romanticism has almost defined itself as being in hostile opposition to rationalism.

SCIENTIST: I am getting thoroughly confused, especially with that infernal din going on outside.

PHILOSOPHER: Let me put it like this. One reason why many people find conventional rationalism objectionable is because of its dishonesty, its hypocrisy. Those who uphold the value, the central importance of reason do not consider themselves to be advocating a philosophy of life. And yet this is precisely what they are doing, but in a highly surreptitious, dishonest fashion. Those who ostensibly object to "rationalism" are really objecting to the way of life, the values and ideals, being advocated by rationalists; and, in addition, there is perhaps the unvoiced criticism that it is thoroughly objectionable to advocate a certain way of life while all the time pretending that one is doing no such thing.

The vital point about aim-oriented rationalism, however, is that it amounts to an ideal for reason which ought to be of use, of value, within *all* philosophies of life. In advocating aim-oriented rationalism, I am not (I hope) surreptitiously advocating a specific way of life: rather, I am putting forward a methodology for developing philosophies of life, and thus lives, which has, I believe, *universal* relevance and application, whatever one's philosophy of life may happen to be.

SCIENTIST: I am sorry. I really don't understand what you're talking about. For one thing, I can scarcely hear what you are saying –

[And just at this moment, the hubbub outside rises to a crescendo, and quite suddenly the door is flung open and into the room there bursts a number of new puppets, all jostling each other, shouting each other down, clamouring for attention.

But what a strange collection of people! The Scientist and the Philosopher can scarcely believe their eyes. There is a Buddhist monk, dressed in saffron robes – or at least someone who looks just as if he is a Buddhist monk. He is smiling ecstatically upon all the confusion as if it is all but an aspect of the One, ultimate reality, the seventh heaven, Nirvana, or whatever. There is a youngish middle-aged man, stern, somewhat humourless, but clearly of good will, dressed in curiously formal, old-fashioned clothes (Rationalist). He looks exactly like the film director, Truffaut, as he appears in his wonderful film, *L'Enfant Sauvage*. There is a young woman, attractive and radiant, who seems to find it all a great joke (Romantic). There is a smallish, fierce-looking man, wiry, red hair, red beard (Rebellious Romantic). His face reminds one vaguely of van Gogh. He seems to be in a state of extreme rage. There is a parson, with a long, white, troubled, somewhat ascetic face (Christian). There is a rather earnest, eager-looking middle-aged man, who gives the appearance of someone who is anxious to please (Liberal). He might be a schoolmaster. There is a young man, wearing small wire-rimmed spectacles; he has short hair and the appearance of someone who is impatient with the whole proceedings (Marxist). Finally, there is a tramp, clutching a bottle of cider, dressed in a long, filthy mackintosh, who is grinning away and muttering to himself (Wino). He goes up to the others, one by one, trying to get himself noticed. Could he be asking for money for a cup of tea? He then spots a comfortable armchair in the corner of the room. Off he goes, curls himself up, and promptly goes to sleep, forgotten by everyone else.

Who are these people? They are not exactly the kind of people one would expect to find on a University precinct. Could it all be a student stunt? Could they be members of the dramatic society? The Philosopher suspects that something like this may be the explanation for this sudden extraordinary intrusion.]

PHILOSOPHER: What on earth is going on? What do you want? [Babble of voices. "We are horrified, horrified at this idea of developing a *rational* society." "Scientific love indeed: is this the sort of thing that goes on in Universities these days?" "At last, someone's taking the whole thing seriously. Only Reason can guide us in these troubled —" "The whole thing is horrifying, horrifying." "Except that it is absurd."] Our only hope of making sense of what it is you have to say lies, I think, in having you speak one by one. Buddhist, perhaps you would begin.

[The seven puppets now expound their philosophies of life, and their attitude towards the suggestion that it is desirable to build a rational society, live a rational life. It must be remembered that our *dramatis personae* are but puppets, capable only, as it were, of expounding puppet versions of the philosophies they claim to espouse. The reader is asked to try putting his own real, vivid, rich, complex, profound philosophy of life into the discussion at this point, to see how it would affect the subsequent argument. Treat the argument as a sort of sausage machine for philosophies of life. If your own philosophy of life gets stuck in the machine somewhere, then bend the machine, the argument, until it lets your philosophy through. If what emerges at the other end doesn't please you, then without further hesitation, throw the argument away as useless.

There is a silence, as everyone waits for the Buddhist to speak. He seems reluctant to leave his blissful contemplation of reality for the realm of action; however, at last, he begins to speak.]

BUDDHIST: We Buddhists believe that, in order to find peace, fulfilment, we need to give up all earthly desires and ambitions. Life lived in pursuit of worldly aims is inevitably doomed to frustration, disappointment and bitterness. Only by giving up all desires, all ambitions, all aims, can one achieve Nirvana, unity with the One, the All. But in order to achieve this, one must even give up the desire to achieve Nirvana. It cannot be achieved: it is granted, when one no longer seeks. And through the practice of meditation one can prepare the mind –

CHRISTIAN: (Interrupting) It is quite clear that there is here the expression of deep religious feeling. But I am afraid our Buddhist friend has not heard the truly marvellous news: reality is God; and God is a God of love. Our task in life is ideally to abandon all earthly desires and ambitions, just as our Buddhist friend says, but for a *purpose*: so that we may do the will of God. Our aim in life should be to submit ourselves to the will of God, not out of fear, or

174

out of a desire for reward, but for its own sake, as did our Lord, Jesus Christ, who died for us all on the cross.

It is *faith*, faith in God and in His love for us, His concern for our spiritual welfare, that we need today. Reason without faith cannot take us very far. A "rational" society would, I am afraid, be a God-less society, a society without faith and hope, devoid of all spiritual values. We need to put our trust in God, not in Reason.

PHILOSOPHER: My ideal of a rational life, a rational society, far from excluding faith and love, actually depends upon, and is designed to promote and encourage, faith and trust, faith and love–

RATIONALIST: (Breaking in impatiently, speaking to the Philosopher and pointing to the Christian) Don't listen to him! We have nothing of value to learn from these religious maniacs, these peddlers and dealers in superstition. Look at the kind of rubbish they talk! Nirvana indeed! And in the meantime peasants die of hunger. Love and faith! Is that what prompted the crusades, the Inquisition? In essence, the thing is very simple. Despite what our Christian "friend" here says, everything that is of value in life does depend in the end on reason. It is to reason that we owe our humanity, our civilization. Without reason, we are mere beasts, motivated by all kinds of violent, destructive passions, or dominated by all kinds of absurd superstitions and beliefs. Only reason enables us to discover truth, develop humane, civilized, rational laws, moral precepts, institutions, modes of conduct. We must use reason to curb our violent animal passions, and our innate propensity to believe all kinds of absurd superstitious nonsense. We must build our life, our civilization, based on the firm assurance of reason herself. I, for one, heartily welcome this clarion call to build a rational society, based on the methods of science.

PHILOSOPHER: (In an aside to the Scientist) Our Rationalist comes, of course, from the 17th or 18th century. Notice incidentally how closely the Rationalist's pattern of thought mirrors the Christian's pattern of thought. Indeed, Rationalism might almost be said to be Christianity, with the meaning of a few key terms shifted, but the basic overall pattern retained. God has become Reason; the soul has become the mind or consciousness; salvation has become knowledge; virtue has become thinking and acting in accordance with the edicts and principles of reason, rigour, scientific propriety; grace is truth, sin error; sainthood is genius; evil temptations of the devil have become irrational passions, desires, impulses, prejudices, superstitions, which need to be curbed, not by appeals to God, but by appeals to Reason. The Church has become the University, the Academic Establishment, the institutional enterprise of Science, tended by its priests. Science, the supremely successful enterprise of Rationalism, can be conceived of as man's endeavour to become God-like, to acquire knowledge, power, mastery over Nature.

Of course, these analogies between Christianity and Rationalism in no way indicate that Rationalism does not deserve to be taken seriously, on its own count, as an independent philosophy of life. It is, perhaps, worth remembering, however, that Rationalism did start its life off – if we leave aside here consideration of the ancient Greeks – in close association with Christianity. Thus, for Kepler, natural philosophy amounted to a religious quest, in that, as a result of pursuing natural philosophy, he hoped to discover, to divine, the thoughts of God embodied in the simple mathematical laws governing the motion of the heavenly bodies. Again, for Descartes, Reason could not, as it were, generate its own authority, its own legitimacy, but could only acquire this from God: faith in Reason would only come from faith in the good will of God. Even for the great Newton, natural philosophy, mathematics and theology, were all intermingled, but in a way that was much more confused than with Kepler or Descartes.

Gradually, however, the immense success of natural philosophy and mathematics led Rationalists to break away from their Christian origins, and to establish Rationalism quite independently of Christianity, and even sometimes in hostile opposition to Christianity. And this, of course, meant that Rationalists could no longer very easily scrutinize their Christian inheritance – in order to see whether they had inherited all that was desirable, and left behind all that was undesirable. If the break had been less decisively established, it might have been easier for Rationalists to

176

acknowledge the value of *faith*, of *trust*, and the value of putting first the interests of those whose need is the greatest, the most desperate. And again, it might have been easier for Rationalists to recognize the authoritarian element in their concept of Reason, inherited from Christianity, but at odds with the deepest intentions of Rationalism, which were essentially democratic and anti-authoritarian.

RATIONALIST: Do not imagine I have not been able to hear those muttered comments. I have heard every word. And the whole thing is a rigmarole of rubbish. Rationalism is on an entirely different footing from Christianity or Buddhism, or any other superstitious creed, religion, or "philosophy of life". Look at our achievements! There is science. Who can doubt but that for the first time in the history of mankind, science is steadily building up authentic *knowledge* about Nature, which is having a tremendous impact for the good as far as the welfare of humanity is concerned. Look at our achievement in mathematics, in engineering, in navigation, in agriculture. This is not some superstitious creed, resting on *faith*. We do not sit around praying ineffectively. We are men of *action*. For the first time ever, mankind is acquiring the capacity to dominate Nature, exploit her ways for the benefit of humanity. You betray our high cause with this whispering about Rationalism being somehow closely associated with Christianity.

REBELLIOUS ROMANTIC: (Unable to contain himself any longer) This man [Rationalist] is a menace. He thinks he has the answer to all our problems. And he is much closer to the cause of all our problems. Everything of value in life depends on Reason indeed. What rubbish! What absolute rubbish! It is feeling, passion, instinct, intuition, imagination, that is the source of all that is of value in life. Life comes from our heart, here, from the quick of our being (and here the Rebellious Romantic thumps himself repeatedly on the chest). Sympathy, love, passion, ecstasy, beauty, wonder, joy, compassion, laughter, delight – it all comes from here, from our hearts. And why do we not live openly, spontaneously and joyfully from our hearts, in common natural friendship? Because of him! Because of that madman there, with his cold, repressive reason, his seething cauldron of repressed

177

hatred, his deathlike inhumanity concealed beneath noble declarations of his concern for the "welfare of humanity". He is terrified! Terrified of himself. Terrified of the vastness of his soul, bequeathed to him by Nature. Terrified of the violence, the impulsiveness, the passionateness of his emotions.

And so what does he do? He invents this horror upon horror, his precious Reason, to curb and repress almost all of his humanity, almost all that is of natural value in his makeup. He thinks it is reason that rules his life. What a joke. It is fear, fear, fear. His precious reason is but the outward expression of his panic. He is like a man so terrified of his murderous impulses that voluntarily he chains himself up, locks the chains around him, and throws away the key. His reason is but chains, imprisonment, compulsive self-inflicted constraint, all arising from the terror of freedom.

And where does all this terror come from? Why is our poor Rationalist so terrified of himself, of his own violence? For that is how it is with him: he is terrified of his own violence, his own hatred, which dimly he divines within his own nature – hence all this talk about beasts in such tones of disgust. What is so terrible about animals? We *are* animals. And animals can be *beautiful*: they play, they run fast as the wind, they soar in the sky: they are free! But our madman here is not free; he is in chains. *And that is why there is such violent hatred within him, such rage, such fury!* He longs, passionately, to be free: but he cannot be free because he is in chains; and so a fury builds up within him. And this makes it all the more imperative that the chains be kept tightly locked about his body: which, in turn, enrages him all the more: and so on, until he goes mad. His reason creates his fury; and then he needs his reason all the more to contain his fury; which, in turn, only enslaves him all the more to his reason. The whole of life, the whole of civilization must be based on fiercely constraining, repressive chains of "reason", so that our "civilized" person will not run amok. He must be caged in by laws, customs, codes of conduct, institutions, all based on repressive "reason". No wonder the "civilized" person becomes world weary, overcome with listlessness, apathy, guilt, devoid of capacity for spontaneity of feeling, for joy, haunted by intimations of brutality, ugliness and madness. And no wonder his precious science should unleash such horrors upon the world: with such a mad pressure of unacknowledged hatred in his heart, what else would one expect?

And what is to be done? It is so simple, so very simple. Overthrow Reason! Overthrow the intellect, the mind, the cold, hard repressive voice of reason, science and logic. Rebel! Let loose all the torrent of impulses choked up in the heart, however violent these may be at first, for then one will discover, beyond all the rage, a miracle of gentleness, grace, beauty, freedom and fulfilment. Unfurl passionate desire! "The road of excess leads to the palace of wisdom" as Blake says somewhere. And speak from your heart to your fellow human beings, from heart to heart. Let the profound, beautiful, miraculous impulses of your heart come out into the world, so that others may see that wonders come from rebellion, and they too will gather up the courage to overthrow repressive Reason, and learn to live instinctively and impulsively from the heart, with openness, honesty, tenderness, *feeling*, for their fellow human beings. It is not just I who says all this. There are so many of us. Blake, Wordsworth, Coleridge, Keats, Hazlitt, Hölderlin, Baudelaire, Rimbaud, Rilke, Rousseau, D. H. Lawrence, Schopenhauer, Nietzsche, Dostoevsky, Kierkegaard, Sartre, R. D. Laing, Roszak, Doris Lessing, and oh, many, many others. It is especially in *art* that we can explore and share the impulses, the feelings and desires, the visions of our hearts. It is not *science* that gives us knowledge, but *art*. It is in art that we can let loose our passionate longings, express freely the soarings and plungings of our impulses. Through art we speak to one another, from heart to heart. Up to a point, yes, the tools of the enemy have to be used – language grammatically constructed, sounds ordered in accordance with the laws of perspective and colour harmony. But, just because they do belong to the enemy, these tools, as far as possible, must be confounded, abused, torn from their customary rational use and shape, and employed to articulate the passionate, irrational, rebellious impulses of our hearts. It is just by overthrowing established conventions of symmetry, balance, proportion, good tastes, rational aesthetic creed, and letting loose a cacophony of colour, a riot of sound, a blaze of reason-defying imagery and thought, that we succeed in avoiding the great lie and revealing the

179

real, repressed truth. It is the frenzied, ecstatic poet, the delirious artist, who tells us the truth, not the scientist, the philosopher, the logician, the academic, the scholar, the cold, repressed, secretly revengeful man of Reason. For the Rationalist, the ultimate tragedy in life is to go mad, for madness amounts to the unseating of Reason, and thus the loss of humanity, a return to mere beasthood. But for us, madness is ultimate wisdom, a visionary experience of reality, a release from the chains of repression, emergence from the dark cave of shadows into the land of reality, illuminated brightly, brightly, by the sun. If we are to be free, if we are to realise our humanity, attain the passionate longings of our hearts, and live with one another openly, honestly and lovingly, as human beings are meant to live with one another, then we must go mad: we must overthrow all these chains constructed so laboriously by Rationalism – all these repressive institutions, systems of government, laws, police, paralysing conventions and customs, and live spontaneously, from the heart. And we must teach all this to our children, encourage them to hold on to their spontaneity, their freedom, their capacity for instinctive self-expression, instead of "educating" them, "training" them rather, to become like us, trapped, world-weary, hate-burdened, chained adults.

If one man knew all this, then surely that man was Blake. Listen to this – listen!

> Tyger tyger, burning bright,
> In the forests of the night;
> What immortal hand or eye,
> Could frame thy fearful symmetry?
>
> In what distant deeps or skies,
> Burnt the fire of thine eyes?
> On what wings dare he aspire?
> What the hand, dare seize the fire?
>
> And what shoulder, and what art,
> Could twist the sinews of thy heart?
> And when thy heart began to beat,
> What dread hand? and what dread feet?

What the hammer? What the chain,
In what furnace was thy brain?
What the anvil? What dread grasp,
Dare its deadly terrors clasp?

When the stars threw down their spears
And water'd heaven with their tears:
Did he smile his work to see?
Did he who made the Lamb make thee?

Tyger Tyger burning bright,
In the forests of the night:
What immortal hand or eye,
Dare frame thy fearful symmetry?

There! When has all your philosophy, your science, your precious reason, ever come up with a thing like that? There are no dead words there. It lives. It burns in the mind. And just try to make "rational sense" of it: you cannot. That third stanza: we begin with questions about God, which imperceptibly slide into being questions about the *Tyger*, his dread feet. It does not make *grammatical* sense: but it makes sense here, here (and again he thumps his heart).

PHILOSOPHER: (Murmuring) I would have thought Blake's poem makes perfect sense. Blake wants to *show* God fusing into the Tyger through the intensity of creation; and just that is arranged by the grammar. The Tyger is God in the world, terrifying because of the ferocity of loving creativity that it represents. You, yourself, O Rebellious Romantic! But the *poem* itself is a very skilfully, intelligently shaped piece of work.

REBELLIOUS ROMANTIC: (In agony) How can you? How can you? In fact, quite generally, everything that you stand for, Philosopher, is unforgivable. That man there (pointing to the Rationalist) is a throwback, an anachronism, having no relevance whatsoever to our times today. *But you!* How can you resurrect the absurd horrifying nightmare of a *rational* life, a *rational* society?

181

Do you want to turn us into ants, into robots, into zombies? We human beings refuse to take part in your obscene, hate-driven plan to destroy all life and turn us into the cogs of your "rational society".

PHILOSOPHER: But my vision of a "rational society" is almost the same as your vision of a society in which people "live with one another openly, honestly and lovingly, from heart to heart".

REBELLIOUS ROMANTIC: *What?*

PHILOSOPHER: I have considerable sympathy with your outburst against our 18th century Rationalist. In my view you make only one mistake. You get a little too carried away with your passion: and that prevents you from realizing that what you are up against is not "Reason" at all, but an irrational concept of Reason, an irrational ideal for reason. You are yourself a *victim* of the very thing you condemn. For, absurdly, you concede the Rationalist his main point: if "Reason" means anything it means what he, the Rationalist, says it means. It is because you concede to the Rationalist this basic point that you are obliged to regard yourself as an opponent of Reason, an Irrationalist. And, in fact, the thing is all the other way round. If anything, Reason is on *your* side.

REBELLIOUS ROMANTIC: What? What? How dare you! I will have nothing to do with it! I wouldn't touch Reason with a barge pole.

PHILOSOPHER: But what you don't want to touch with a barge pole is an *irrational* concept of Reason, not Reason herself. Reason herself is actually what you believe in most passionately, what stands at the very centre of your philosophy of life.

REBELLIOUS ROMANTIC: (His hands over his ears) I just cannot take any more of this. How dare you desecrate everything that I most value, that I hold most precious?

PHILOSOPHER: But you are just reacting emotionally to *words*, ignoring *ideas*. You, yourself, said it. "Unfurl the passionate desires of your heart in your imagination".

REBELLIOUS ROMANTIC: Yes!

PHILOSOPHER: Well that constitutes the heart of a truly rigorous ideal for Reason. Unfurling our passionate desires in our imagination is the *essence* of Reason. And in defending this against "Rationalism", so called, you have been defending Reason against Irrationalists. One does not become rational just by *calling* oneself a rationalist, you know.

REBELLIOUS ROMANTIC: I cannot bear it. I cannot bear it.

PHILOSOPHER: But if you calm down for a moment, and just consider the thing quietly, in tranquillity, I am sure that you will be delighted with what I have just said.

REBELLIOUS ROMANTIC: Don't tell me to calm down! Do you want me to fall into despair? How can I hope to keep going if I do not rage, rage, against all the deathly calm murdering of all that is good and gentle and warm-hearted and human in life?

PHILOSOPHER: (Quietly, in an aside to the Scientist) Incidentally, I don't know if it has occurred to you, but one can regard the ideas of Freud as an attempt to develop some kind of compromise position between these two extremes of "Rationalism" and "Romanticism". According to Freud, civilization places the most appalling constraints upon the Id, the seething ocean of our tumultuous desires. Growing up is a process of repression after repression, a process of frustrating instinctive desires, and then burying from oneself all knowledge of these desires. Impulses of love – for one's mother, perhaps – lead to impulses of murderous hatred – for one's father; in an attempt to cope with these appalling impulses, we bury them, together with the impulses of love from which they sprang. But they continue to direct our actions, in a secret and subversive fashion. Our Reason is but rationalization. Far from our Reason controlling our passions, our repressed passions control our Reason – the ultimate confounding of the Rationalist's hopes. So far, Freud is all on the side of Romanticism. But then Freud goes over to the side of Rationalism. Without repression, civilization would be impossible. The almost total frustration of the Id's desires is essential if civilization is to exist. Misery, neurosis, the failure to realise our deepest, most passionate desires, is the price we pay for civilization. The most that we can hope for is to achieve *symbolic* gratification of our deepest desires – via sublimation. Neurosis is incurable, happiness impossible. The best that psychoanalysis can offer is to help us to discover how we can live with our neurosis, bear the burden of our inevitable misery. Thus Freud sternly upholds the basic ideals of traditional "Rationalism" – and, of course, his whole work was contributed through the vehicle of *science*, carefully distanced from his own personal feelings, desires and values.

REBELLIOUS ROMANTIC: Ah, Freud. He understood a thing or two about the human predicament, and the delusions that people are prone to – especially the glib facility of "Rationalists", so called, to indulge in *rationalization*.

SCIENTIST: Freud's ideas may have, for some, a certain imaginative appeal. But as an intended contribution to *science*, Freudian theory does not, I am afraid, even begin to satisfy the exacting standards of scientific testability and experimental verification demanded by science. Freud may, for all I know, have made a contribution to literature, or to metaphysics; but he has not, I am afraid, made a contribution to *science*.

PHILOSOPHER: Do not be so sure Freud can be dismissed so lightly. Our methodological, rationalistic reinterpretation of Freud is going to upset very seriously your airy dismissal of his ideas. It will turn out that what really needs to be said is that *science* itself fails to comply, quite drastically, with elementary requirements of rationality and rigour demanded by *methodological psychoanalytic theory*. You see, Rebellious Romantic, I really am, to a considerable extent, on your side!

SCIENTIST: (Laughing) Oh yes! I had forgotten your obsession about science being *neurotic*.

PHILOSOPHER: Who is next, then, to tell us what really matters in life, and why the ideal of a rational life, a rational society, would inevitably blot out all that is of value in life?

(NON-REBELLIOUS) ROMANTIC: (After a silence) I suppose it must be my turn.

PHILOSOPHER: Off you go then. Tell us all how to live.

184

ROMANTIC: (Laughs) That's the last thing that I would want to do!

PHILOSOPHER: At last! Someone who is actually *reluctant* to tell others how to live. What a refreshing change.

ROMANTIC: Well, for a start, I don't think I have a fully-fledged "philosophy of life" as everyone else here seems to have. I'm certainly not a "Romantic": I can't think what induces you to call me that. My life is much too prosaic and full of humdrum practicalities to be "Romantic". I think I would like to be a *Realist*: but I don't think I am in practice very realistic about things. It's very easy for me to get things out of proportion. With the slightest encouragement, or even with no encouragement at all, I imagine that the worst will happen. But basically I am too busy bringing up my children to have the time or the energy to work out my "philosophy of life". You've got the wrong person I'm afraid (and she laughs: she doesn't seem to be put out at all by the fact that she is not going to be able to help on the discussion). And in any case, I have far too many problems trying to work out how to live myself to go around telling other people how *they* ought to live.

PHILOSOPHER: Why don't you just tell us what seems to be to you the most important thing in life. Never mind about the question of whether it is important for *others* as well.

ROMANTIC: What do I think is the important thing in life? (A long, reflective silence) Love.

PHILOSOPHER: And what is love?

ROMANTIC: If I had known I was going to be subjected to an examination – and an examination on *Love* of all things – I would never have agreed to come here in the first place.

PHILOSOPHER: Never mind. It isn't an examination. Just say a few stupid things that occur to you, on the instant, and we'll go on to the next person. Obviously you can't tell us what you *really* think, just like that, in this highly artificial situation. Not for a moment will any of us make the mistake of thinking that what you say now represents what you *really* believe. It's just something to throw into the discussion for a laugh, to keep it going.

185

ROMANTIC: Well, I'm not at all sure that I believe in "romantic love" for a start! Let me see. Love is a feeling one has for another person, a feeling for the value, the preciousness, of the other person. But it isn't just a feeling: it involves actively caring for the person you love, making things go well. It involves sharing one's life, one's self, I can't imagine real love without honesty, saying what one really feels, even if at times what one feels is all wrong. I expect anger, breakdown of communication, go on even between people who have a very loving relationship with each other. This is all sounding very stupid, but never mind: you asked for it! I think it is important to be *honest* about one's feelings and motivations, whatever they may be. And to love the other person's reality, what they really are, not one's own image of what they are. We are probably all inclined to *use* the people we love for our own purposes, rather than being for both our purposes and their purposes. It's all very different from what you *think* it's going to be like. I think people are often quite *frightened* of love: perhaps people are frightened of intimacy, I don't know.

I must say, I don't think reason has much relevance to any of this. I think one has to rely to some extent on one's feelings, one's instincts, one's impressions, of what is going on. A person who thought reason was relevant to love would probably be just trying to *bully* the other person, get the other person to do what *they* wanted him or her to do, under the subterfuge of it being the "rational", the "logical" thing to do. I suppose I am inclined to think that people who believe in logic and reason and argument are often trying to escape from their feelings. It seems to me to be better openly to acknowledge what one feels and wants, rather than to dress it up, disguise it, as an "objective argument". At least you know where you are with a person who is not afraid of saying what he wants, and what he feels. Though of course "being honest" about feelings is open to all kinds of abuses.

LIBERAL: We are, I am sure, all very grateful for your simple and sincere words on such a difficult and personal subject.

One point does, however, occur to me. It seems to me that it is quite possible to combine elements of Rationalism and Romantic. ism into a coherent whole, without necessarily being obliged to accept the views of Freud. You see, in my view, what is of supreme value in life is *people*, individual human beings. Society is for people, not people for society. We need to construct our institutions, our economic, industrial, legal and political systems so as to promote individual liberty as far as possible. And all this requires careful thought, planning, intelligence, reason. In addition, it is perfectly clear that science and technology can be of immense value to humanity – and science and technology clearly depend upon rationalistic capacities and skills. Within the public, objective, factual domain, reason, science, knowledge, logic, Rationalism are all important.

But the realm of public, objective fact does not comprise the whole of life, the whole of reality: there is also the realm of subjective feeling and desire, subjective values, imagination, art, beauty, kindness, love, charity, friendship, personal fulfilment – all that *Romanticism* has traditionally concerned itself with. It is within this personal, subjective domain that we find meaning, significance, value. In the end, perhaps, it is this personal, subjective realm that ultimately matters.

The all important thing is to strike the right balance, maintain a proper sense of proportion, between these two domains. We need to give support to Rationalism within its own sphere of application – the objective, factual, public world. And we need to give support to Romanticism within its own sphere of application – the world of emotion, values, imagination, personal relationships. Science, logic, scholarship, explore, and provide us with valuable knowledge of the objective, public, factual territory. Art, music, literature, poetry, explore the interior, personal, subjective world, and provide us with valuable knowledge and insight concerning this personal world. Both are needed: both are of value. Things only begin to go wrong when the methods, the skills, the knowledge, the values of one domain start being applied where they are utterly inappropriate – namely, in the *other* domain. Our Rationalist is quite right to value science, reason, logic, scholarship; he is quite *wrong*, however, to imagine that Reason is appropriate and helpful for the whole of life, the whole of reality. Our Rebellious Romantic is entirely justified in protesting so fiercely at the illegitimate intrusion of Reason into the personal, subjective world of emotion, desire, imagination and art. But he begins to go disastrously wrong when he gives no place to the Rationalist and to the value of Science and Reason. When he begins to deny the existence of the objective, public, factual world altogether – as if the whole thing were a conspiracy, an hallucination, of Rationalism – then he does indeed begin to advocate madness; he almost becomes mad himself. A figure of speech, you understand: I don't mean anything *personal* by that at all, of course. (Here the Liberal smiles anxiously at the Rebellious Romantic, but is ignored.)

The all important thing, then, is to keep a sense of proportion, a proper balance between the two extreme, unacceptable positions. In education, in politics, in psychiatry, in a host of human activities and enterprises, we need, to define the boundaries of the two territories clearly, and give proper, just support to both. Thus, in education – which happens to be my own particular concern – it is important that we give due emphasis to academic learning, to the acquisition of essential intellectual skills: but in addition it is important that we give due emphasis to self-expression, to opportunities to exercise the imagination. Both aspects of education need to be emphasized: and education suffers when one aspect begins to dominate at the expense of the other.

PHILOSOPHER: (Murmuring to himself) All in the best spirit of British compromise, in fact.

LIBERAL: What? I'm not quite sure I heard that. But to conclude: as you will have been able to gather, Philosopher, from what I have been saying, I can only react negatively to your suggestion that it is desirable to *live* rationally, to *love* rationally, to develop a rational society. Your suggestion amounts to a massive take-over bid by Rationalism into territory where it has no right to be, and can only do harm, namely the territory of Romanticism. Science is one thing; art is another thing. Thought, mind, intellect, logic, reason, fact, objectivity, all have their place, their value: and feelings, desires, aspirations, imagination, values, ideals all have their proper place, their value. We need to *preserve* these distinctions.

There! How does that sound? What are your reactions to that?

PHILOSOPHER: You have, I think, articulated with great clarity a viewpoint that is widely upheld these days. Indeed the viewpoint might almost be said to correspond accurately to the *reality* that we have created. We do tend to find the world divided up into two distinct domains, just as you have described. On the one hand, there is the public, objective, factual world, to which science, technology, scholarship, knowledge, intellect, mind, reason, are held to be appropriate. On the other hand there is the world of personal, subjective experience, the private world of feelings, desires, dreams, values, to which art is held to be appropriate. And the two kinds of world often seem to be somewhat dissociated from one another, in only somewhat distant, ineffective communication with one another.

LIBERAL: It sounds as if you find the whole thing somewhat undesirable.

PHILOSOPHER: Yes, I do find it undesirable. In reality, after all, there is but *one* world. Splitting our one world up into a common, public, objective, factual world, and all our distinct inner, private, personal, worlds of subjective experiences and values seems to me to be something like the underlying cause of a great number of our troubles today. We all *ought* to be thinking, I believe, in a quite straightforward way, about relationships between people and cosmos and people and people – as I have emphasized a number of times already, I am afraid. The kind of scheme that you have outlined for us, Liberal, disrupts precisely such straightforward thinking. Our thinking about our *personal worlds* falls into one category; our thinking about our common public world falls into another category. Thus, we fail to think straightforwardly and intelligently about the interactions between the public and the personal, the objective and the subjective – which, of course, is to fail to think straightforwardly and intelligently about relationships between people conducted through the public, objective world, as it were. We receive no encouragement to take *personal responsibility* for the so-called public, objective world: indeed that tends to seem an almost lunatic undertaking. If one really did assume a sense of personal responsibility for our common, public world, then I think most of us here would find it very difficult not to fall into a state of *despair* – partly because of the really very grim, horrible, insane aspects of our common public world, of which we are all aware, partly because of the feeling of almost complete personal impotence and powerlessness that would overwhelm us. Our state would not be thought of as an entirely natural, sane, realistic, sensible reaction to public reality: on the contrary, we would be held to be suffering from a serious mental condition – clinical depression. We, ourselves, would be inclined to think of our own state in such terms. The kind of split that you are advocating, Liberal, in our thinking about objective, public reality, and personal, subjective reality, makes it insane, clinically insane, to connect up the two worlds, assume a share of *personal* responsibility for the public, objective world we have created and are creating.

But if only the *insane* assume a share of personal responsibility for our common, public objective world how can we expect our common world to evolve in humanly desirable directions? If we do not assume sane, adult, steady, balanced responsibility, openly and publicly, as a part of our personal life, how can we expect our world to become more humane? Is it not clear that it can only be by looking at the inter-relations between the objective and the subjective, the public and the private, the impersonal and the personal, facts and what we desire and value, that we can hope to create a world that begins to correspond to that which we know to be humanly desirable? Dissociate our objective, public thinking and acting from our personal, subjective thinking, feeling and desiring, and the objective, public world will develop in ignorance of our personal feelings, desires, values.

In short, in my opinion, the viewpoint that you have expounded for us fails to go to the heart of the problem. It fails to resolve the clash between Rationalism and Romanticism. It is an unsatisfactory compromise position, not a resolution of the conflict. You see, you are quite wrong in suggesting that what I am proposing is a massive invasion of Rationalism into Romantic territory. What I have to suggest amounts just as much to an infusion of Romanticism into Rationalism, as to an infusion of

Rationalism into Romanticism. What I have to propose is, I believe, a genuine synthesis of Rationalism and Romanticism, in that it is an improvement over conventional Rationalism on conventional Rationalistic grounds, and an improvement over conventional Romanticism on conventional Romanticist grounds. Romanticism is absolutely right to emphasize the central importance of emotional and motivational honesty, the central importance of unravelling desire in the imagination: all this lies at the heart of aim-oriented rationality. According to aim-oriented rationalism, reason is desirable simply because reason helps us to discover and realise the truly desirable. Reason is an offshoot of desire – and not at all something in *opposition* to desire, or apart from desire, as conventional Rationalism would have it. At the same time aim-oriented rationalism is demonstrably a more rigorous ideal for reason than conventional conceptions of reason.

MARXIST: (Impatiently breaking in) So far I have been very patient. I deserve I think to get a hearing at least. I will be very brief. If you really want to understand what is going on in the world today, then I suggest you look at the whole system of economic exploitation that exists at present. Those who produce wealth by their labour, the workers, are mercilessly exploited by the capitalists, those who at present own the means of production. That is the elementary and fundamental injustice of our capitalist society. And until this basic exploitation is put right, all the rest is just empty, dishonest talk. The importance of looking at relationships between people in society has been stressed. And yet so far not a word has been said about what is the most glaringly obvious fact about relationships between people in society, namely, that in a capitalist system, man is bound to exploit man – and, er, woman as well. There is a quite fundamental conflict of interests built into the very fabric of our society, between the working class and the capitalists, the bourgeoisie. Inevitably, this built-in class conflict must lead, in one way or another, to the destruction of the capitalist system. Until that happens, all this talk about "aim-oriented rationalism", the "rational society", "person-centred science", is just idealistic hot air, Utopian rubbish, typical liberalistic evasiveness.

PHILOSOPHER: Well, I —

SCIENTIST: Please! At this stage we cannot possibly afford to get ourselves lost in *that* debate. I have been waiting patiently for you, Philosopher, to clarify those remarks you were making about rationalism when we were so abruptly interrupted.

PHILOSOPHER: Yes. It is very simple. One point has I hope emerged with crystal clarity from our discussion: 17th and 18th century Rationalism is just as much a "philosophy of life" as Buddhism, Christianity or Romanticism. The exhortation to uphold Reason, in this sense, is an exhortation to adopt a certain way of life, adopt a certain philosophy of life which might be put like this: submit your ideas, your principles, your values, your actions, to the judgement of Reason.

Already we may feel that there is here a basic dishonesty. Reason, surely, should help us to appraise, choose, and develop philosophies of life; it should not itself *be* a philosophy of life. Rationalists thunder away at us about the importance of Reason; and all the time, covertly, they are trying to get us to adopt certain values, certain aims in life, certain views about life – a philosophy of life. Not very rational, one is inclined to say.

But much worse is to come. The distinctive feature of 17th and 18th century Rationalism, as we have seen, is the authoritarian concept of Reason that it upholds: the ideal is to try to arrange things so that *Reason* can decide things for us as far as possible. Originally, of course, Rationalism was held to be just as relevant for the realm of *values* as for the realm of *facts* – as Spinoza's *Ethics* modelled on Euclid illustrates, and as numerous attempts to put moral principles on a "sound rational footing" bear out. Being a philosophy of life it is of course *inevitable* that Rationalism should deal with values – with what we hold to be desirable in life – as well as with the realm of facts.

With the development of Protestant thought, Romanticism, and Liberal thought, however, it began to seem increasingly obvious to people that in the realm of values *people ought ideally to be free to reach their own decisions, their own judgements.*

An authoritarian ideal for Reason, making decisions for us, as it were, began to seem wholly out of place in the realm of values, wholly undesirable. Rationalism, interpreted as a philosophy of life, thus began to face a severe, though somewhat poorly understood crisis, a severe problem of conflict of interests. It's the same problem, do you remember, Scientist, that we discussed in connection with philosophy? It is clear however that the problem faces the whole of scholarship and academic learning, science, mathematics, the humanities, and so on, in that all this has more or less been spawned by Rationalism. Philosophy is just the worst casualty.

Given this severe crisis confronting Rationalism, there are essentially two paths we may take in an attempt to *solve* the problem.

1. We may continue to hold that Reason has a very real relevance and value for *life*, for philosophies of life, for desires, ideals, aspirations, values. In order to accommodate the new idea, the new value-judgement if you like, that it is important that individual people should be free to reach their own decisions and judgements about how to live, about what is of value in life, we need to develop a new, non-authoritarian, genuinely *universal* ideal for Reason. That is, we need to develop a concept of Reason which has explicitly built into it the idea that the whole purpose of Reason is to help *us* to make the decisions, the choices, the judgements that we want to make, instead of the purpose of Reason being to make decisions, choices, judgements *for* us as far as possible. We want a new person-centred ideal for Reason in other words! We want a concept of Reason designed explicitly to enhance our individual freedom – enhance our capacity to choose wisely and well, make the choices that we really do want to make – as opposed to a concept of Reason designed as far as possible to deprive us of all freedom of choice except for the one single choice: to think, act, live, in accordance with the edicts of Reason. Reason itself must not tell us how to live: we ourselves decide that; and we *use* Reason to help us *improve* on our decisions, our lives, in terms of what we ourselves judge to constitute improvement. In short, Rationality, as interpreted here, must not be restricted in its

application and use to one particular *kind* of philosophy of life or way of life: it must be applicable and valuable within all possible philosophies of life. It must be truly universal.

All this allows us to retain the idea that life, and philosophies of life, lie at the centre of rational intellectual concern. What is a philosophy of life? It is an idea, a view, concerning the aims and methods of life, an idea about what is most desirable, of greatest value, in life, what problems arise in realizing what is most desirable, and what methods can be used to overcome these problems. These are the issues, the concerns, the problems that lie at the heart of rational intellectual enquiry. Science, mathematics, technology, logic, etc. are but offshoots of this central concern. This approach, incidentally, leads to the conclusion that it is as important to have non-authoritarian concepts of reason, science, knowledge, in the realm of facts as it is in the realm of values. The ideal is for us individually to make up for ourselves our lives, our Cosmos – using of course the experiences, the ideas, of others, to help us do this. Or to put the thing a little more accurately, perhaps, the ideal is for us to scrutinize and develop for ourselves the ideas about the world, life, ourselves and others, that we have inherited, in one way or another, from others, as we have grown up. Detailed, technical matters may well be delegated to experts: but the proper rational task for the experts is not to take important decisions *for* us – even decisions about purely factual matters. Rather the rational task of experts is to lay before us in simple, clear, honest, non-technical, terms, the choices that lie before us, and what their consequences are – the very thing, in fact, that I am trying to do at the moment. So much for the first choice.

2. The second option open to Rationalism is to retain the authoritarian ideal for Reason, retain the idea that, ideally, Reason ought, as it were, to take decisions out of our hands, and make them for us. But if we are to adopt this option, and if we hold in addition that *values* ought to be decided upon by people, then we will be obliged to restrict Reason to the realm of *fact*, to deciding questions of value-neutral fact for us, to helping us, in short, to increase our store of *knowledge*. Values, we will be obliged to say, lie outside the domain of application of Reason altogether.

By now you all, I am sure, completely understand my position. In my view Rationalism *ought* to have plumped for the first option. The first option is the truly *desirable* option, the honest option, the option that allows us to develop rational intellectual enquiry in such a way that it is directly, actively, intelligently, sensitively and effectively for *life*, for the flowering of human life, as opposed to being in some ways almost anti-life, or at least obstructing clear, simple, direct, helpful, thought about life.

Disastrously, the Rationalist tradition plumped for the second option – without really very clearly understanding what the *problem* was, and what possible solutions to the problem were available. Plumping for this second option appeared to have, for Rationalists, several advantages. It seemed to make it possible to retain authoritarian concepts of science, knowledge and reason intact. Rationalists were in any case more interested in improving knowledge about the world (or improving knowledge about the realm of mathematical entities) than they were in improving *life*, working out what life is all about. Thus it was, *for them* (not for the rest of us), no great loss to transform Rationalism from a philosophy of life into a philosophy of how to attain knowledge. Retention of the authoritarian, oracular conception of science and reason enabled Rationalists to retain the idea that science and knowledge would gradually obtain for us power over Nature, mastery over Nature – though of course such ideas slipped into the unconscious since they had more to do with Rationalism as a philosophy of life – a religion, as it were – than as a philosophy of knowledge. Rationalism continued of course to be a philosophy of life – but in an unacknowledged, disavowed, *dishonest*, fashion. The new Rationalism was based on the absurd assumption that the most valuable thing in life is knowledge. (Improving *knowledge* took priority over improving *life!*) Thus the main activity, the main goal, in life became: to acquire knowledge. The fundamental *problems* of life became: How can we attain the main goal of life – namely, knowledge? The central philosophy-of-life problems of the new Rationalism had become: the problems of epistemology and methodology! Western philosophy was on its way! The academic life has not looked back ever since. Science, mathematics, philosophy, the humanities, educational studies, engineering, medicine, etc. etc. all spawned by the new Rationalism, all in the thrall of this absurd philosophy of life which declares about itself that it is *not* a philosophy of life. Finally, of course, this philosophy of life of the new Rationalism ends up declaring about itself that it is not any kind of philosophy at all! And this, of course, is *standard empiricism.* The new Rationalism, whittled down to the point of present-day profoundly influential absurdity, is standard empiricism – a philosophy which utterly dominates scientific activity, and at the same time declares that scientific activity is completely uninfluenced by philosophy!

Just a *hint* of how disastrous, how harmful, has been the Rationalist decision to plump for the second option rather than the first, thus apparently retaining the authoritarian, oracular, justificational aspects of science and reason, can perhaps be gathered from the following additional considerations. First, of course, this second option fails to capture what was desired – in that science and reason are not entitled to make authoritarian, oracular, claims to achieve knowledge, indubitable, certain, verified knowledge, as Popper, in particular, has spent his life seeking to emphasise. Second, the whole conception of science and academic, intellectual enquiry, that the new Rationalism is inevitably committed to, once the second option has been taken, *fails* to make rational sense of scientific, intellectual enquiry, and actually serves to obstruct the attainment of just that which the new Rationalists care most about, namely, knowledge and understanding of Nature. *That* of course has been the outcome of our long debate, my friend Scientist – at least in *my* opinion, if not in yours.

There is, however, a *third* and far more devastating outcome that emerges from adopting the above second option. An immediate apparent consequence is that Reason relates only to the discovery and assessment of value-neutral truth. An *idea*, in order to be rationally discussable, has to be an idea about value-neutral truth. *At a stroke all philosophies of life become irredeemably irrational, just because philosophies of life of course are ideas which combine factual and evaluative ideas* —ideas about what we want and what facts prevent us from getting what we want. Relative to the ideal

196

for Reason of the new Rationalism, all philosophies of life are inherently irrational, intellectual rubbish, mere ideology, propaganda, mythical, religious, or political dogma and twaddle.

The one thing that ought to be at the centre of our attention – namely, our lives, and our ideas about our lives – becomes something that is not fit for serious intellectual consideration. It is scarcely conceivable that science and reason could do us a greater disservice than this!

No wonder the new Rationalism has not been able to understand itself and its historical origins. It began life as a philosophy of life: then, as a result of a failure to solve satisfactorily a central problem within this philosophy of life, the conclusion was reached: all philosophies of life are inherently irrational. The very origins of the new Rationalism – which now rules our world – had become profoundly irrational by the very criteria of rationality embodied in the new Rationalism. The link with the past had to be severed!

Loud and clear to the tree tops it should be shouted: Of course philosophies of life can be intellectually respectable. Our philosophies of life are potentially the most precious ideas that we have, just because they are so intimately associated with our lives. (To say this is not of course to say that we should go around trying to ram our own philosophy of life into the minds and hearts of others.) In fact, we should, I believe, say this. At the centre of our concern, our attention, is *life*. Intellectually, this amounts to putting at the centre of our concern *philosophies of life*: in thinking about life we "think" philosophies of life, if you like. Exploration of factual issues – science – is but an offshoot of our concern with philosophies of life, an investigation of the more publicly agreed, objective, factual aspects of our philosophies of life. Epistemologically, philosophies of life take priority over theories about the nature of the universe: that, in essence, is what person-centred science amounts to.

Thus, of course, philosophies of life can be rationally assessed – which is not to assert that from a rational standpoint we all ought to agree, for of course our lives are different, our interests and concerns are different, *we* are different. Diversity is both desirable and entirely rational (and *not* just for epistemological reasons, arising from the fact that we cannot *know*: even if we all had ready access to the "truth", diversity would still be rationally desirable). In addition, our philosophies of life – and so our *lives* – can be rationally developed. But this means that we can use reason in order to help us to choose the philosophy of life we really want to choose. We can use reason in order to help us develop our philosophy of life, and our life, in directions in which we really want to develop them. It is only with respect to seriously defective, unrigorous, authoritarian conceptions of reason that philosophies of life come out as intellectually disreputable and rationally unacceptable.

There, Scientist, Buddhist, Christian, Rebellious Romantic, Non-rebellious Romantic, Liberal and Marxist: Do you understand now a little better why I share all your distrust, your suspicion of "Reason" as conventionally conceived of, as advocated by our Rationalist sitting sullenly in the corner over there, and as advocated in a modified form by our Scientist here? I share all your suspicions: and yet I still say: Reason has a profound relevance to life. For it is unrigorous, irrational, dishonest, ideals for Reason that are undesirably applicable to life. The undesirability of these concepts of Reason is indeed closely connected with the irrationality of these concepts. The moment Reason is made more *rigorous*, however, Reason becomes profoundly desirable, from a human standpoint, whatever one's philosophy of life may be. (No longer does using Reason amount to adopting a certain way of life – the Rationalist's.) Indeed, as we make our ideal for Reason more rigorous, the idea emerges that Reason, rigorously understood, is but that which helps us to discover, and realise, that which we judge to be truly desirable, truly valuable. By definition, one might almost say, Reason is desirable.

So, don't sulk, Rationalist! You really do have something very precious to teach us. If only you would stop trying to *bully* us to become Rationalists, to adopt your particular way of life and set of values, and instead quietly reveal to us the very valuable, desirable idea that you have developed, allowing us to discover for ourselves just how valuable and desirable it is, you would I think find that you are able to share your discovery with others much more effectively and fruitfully.

CHAPTER NINE

AN AIM-ORIENTED IDEAL FOR REASON

SCIENTIST: It seems to me that during all this time, Philosopher, you have only really been making a few preliminary remarks, clearing up a few possible misconceptions, attempting to gain some initial sympathy for the suggestion that you wish to put forward. But we still haven't had the suggestion itself. You keep *talking* about your new "desire oriented" ideal for Reason, more rigorous, you claim, than conventional conceptions of Reason. But you never seem to get round to telling us what it is. And I, for one, am beginning to get a little impatient.

PHILOSOPHER: Yes, I am so sorry, I –

SCIENTIST: Would it be all right if I quickly summarized the account that you have already given me of your proposal, for the benefit of the others who arrived late?

PHILOSOPHER: Yes, of course.

SCIENTIST: Your suggestion, as far as I understand it, amounts to this. You begin by asserting that your concern now is to apply not the *products* of scientific inquiry to life, but rather the *method* of scientific enquiry to life – to all our various personal, institutional, social activities and endeavours. You point out that a highly distinctive feature of science is its capacity to meet with steady – or even accelerating – progressive success, at least on the intellectual plane. This striking characteristic of science is in marked contrast to almost all other human endeavours – art, politics, morality, religion, international relations, our personal lives. Your suggestion is that if we apply a suitable generalization of scientific method to life, to all our various life activities and endeavours, then we will be able to get into *life* the same kind of progressively successful quality, when judged in human terms, that we already find in science, when judged in intellectual terms.

You argue that, so far, it has not been possible to exploit this idea precisely because the methodology inherent in science facilitating progressive success in science, has been disastrously misunderstood. As long as we take *standard empiricism* as our starting point, we shall not be able to arrive at a methodology, a concept of reason, that is very helpfully applicable to all our diverse life pursuits. If, however, we take the more rigorous ideal for science of *aim-oriented empiricism* as our starting point, the whole situation is, in your words, "profoundly and dramatically changed". Aim-oriented empiricism can easily be generalized to form an ideal for reason (which you call *aim-oriented rationalism*) which has, you claim, immensely helpful implications and applications for all our diverse personal, social, institutional pursuits, and can help us to get into life the kind of progressive success found in science. The finest exemplification to be found so far of aim-oriented empiricism in science is just one of the most strikingly successful episodes in modern science — namely Einstein's development of special and general relativity. It is just this strikingly successful Einsteinian methodology of scientific discovery, which, if suitably generalized, can be employed to achieve immense human success in life.

There! Does that do justice to what it is you wish to propose?

PHILOSOPHER: As usual, you have summed up very accurately and succinctly what I –

SCIENTIST: Then do please give us your detailed *argument*.

PHILOSOPHER: Yes, of course. Let us begin with a brief consideration of what is in my view the finest attempt in recent times to defend standard empiricism as a rational, rigorous ideal for science, and as an ideal capable of generalization to form an ideal for reason of universal relevance for life. I have in mind the work of Popper.

According to Popper, the distinctive feature of science is that in science our ideas, our theories, are exposed to a ruthless process of constant attempted experimental refutation. We cannot verify theories. All our knowledge must remain for ever conjectural, speculative. We can, however, falsify theories experimentally and

201

observationally. We can discover error in our theories. And it is this which accounts for the progressive character of science (insofar as such a thing is possible). We can learn from our mistakes. We can move progressively towards the truth by weeding out error. Thus, for Popper, scientific method consists in essence of the method of trial and error. It amounts to proposing bold conjectures which are then submitted to a process of ruthless attempted experimental falsification. In order to qualify as scientific, it is essential that a theory must at least be open to this kind of fierce, objective, empirical appraisal. A theory must at least be experimentally testable.

Popper has made it clear that, in his view, all this amounts to a special case of something more general, namely the method of criticism. Even when our ideas, our proposals are not experimentally falsifiable – and hence are not scientific – we can still seek to criticize them. We can hunt for inconsistencies in our ideas. We can seek to determine whether some new idea conflicts with other ideas that we may wish conjecturally to accept. Above all, we can try to determine whether a new idea really does solve the problems that it was intended to solve. Reason is thus in essence this kind of *critical* approach to ideas, proposals, proposed solutions to problems. Rational enquiry amounts to proposing bold conjectures controlled by fierce criticism. Rational action involves the critical scrutiny of ideas and proposals that are implicit in action. And science simply represents a special case of all this: in science we have available an especially decisive and objective form of criticism – namely, experimental refutation. An experiment, an observation, is in essence an *argument*, a criticism.

All this is, in my view, important, a part of the truth, and a distinct improvement over traditional Rationalist ideas of scientific method and reason. Rationality, for Popper, is clearly anti-authoritarian, and anti-dogmatic. Scepticism, instead of being the great enemy, which the Rationalist must seek to banish, becomes instead the core of Reason. Again, Popper's ideas of science and reason give an essential rationalistic role to imagination, since without imagination we could not be able to think up bold conjectures to test and to criticize. Even testing and criticism involve imagination; often in order to test severely, or criticize, a theory we need to think up some rival theory. Popper's conception of Reason is thus several steps towards something that might appeal to Romantics, when compared with the ideas of the Rationalists of old.

And yet we may have our doubts. Romanticist, would you like to give us your reactions to this critical conception of reason advocated by Popper? Tell us in particular what seem to you to be its shortcomings as far as the kind of things you are especially interested in are concerned.

ROMANTIC: But that is absurd. How can I be expected to comment intelligently on a piece of philosophy when I don't know anything about philosophy?

PHILOSOPHER: Never mind about that. Just tell us what your reactions would be if someone suggested to you that the critical approach is what is needed in order to solve your problems, successfully achieve what you really want to achieve.

ROMANTIC: A first reaction is, I suppose that it all sounds highly *intellectual*. Reason seems to be, for your Popper, applicable in the first instance to *ideas* and *theories*, and only indirectly applicable to feelings, desires, *actions*. The critical attitude is clearly important; but other attitudes are also important, and help us to achieve success. Sometimes an attitude of indulgent and sympathetic support and encouragement is as important as criticism. We do, on occasion, need to do things spontaneously, and with confidence and certainty. New efforts, imperfect first attempts, may need to be encouraged rather than criticized, if they are to grow and develop. There can be something rather negative and discouraging about incessant criticism and doubt. Creative people, in the arts for example, often do not regard criticism as especially helpful. Sympathetic suggestions for possible fruitful lines of development and improvement can sometimes be more helpful than negative criticism. We learn surely at least as much, if not more, from our successes, our achievements, as from our failures, our mistakes. In the end, it seems, Popper does assign to Reason something of the restraining, repressing, controlling function that was given to it by old-style Rationalism. Emotion, passion, desire still need it seems to be rigorously controlled by Reason in terms of Popper's viewpoint.

PHILOSOPHER: What you say is, I think, absolutely correct. Popper, in fact, took as his basic problem: How can we distinguish *genuine* science from pseudo science? What entitles us to keep the cranks out of science? The problem is formulated negatively and repressively. Popper does not take as the *fundamental* problem: How can we achieve success? How can we, perhaps, be even *more* successful than at present? Science and Reason are conceived of negatively rather than positively and helpfully. And that has, I believe, a great deal to do with the fact that Popper is unable to give to Reason any role in helping us to discover, to invent, to create. The creation of our conjectures, our trials, is in essence for Popper an irrational process: rationality comes in only with the *assessment* of what irrationality has produced.

You will be amused to hear, however, Romanticist, that Popper himself would dismiss your criticisms out of hand, without even bothering to reply to them in detail.

ROMANTIC: Why?

PHILOSOPHER: He would say that you were mixing up rationalistic considerations on the one hand, and emotional, psychological considerations on the other hand.

ROMANTIC: How silly!

PHILOSOPHER: Yes. It shows that criticism, in order to go home, has to be very delicately contrived in order not to be dismissed as irrelevant. However, we, for the moment, can agree amongst ourselves, as it were, that applying the *methods* of science, generalized in the way suggested by Popper, to the pursuits and problems of life, would not in practice enable us to get into life the kind of progressive success that we find in science. Criticism is, after all, a fairly widespread feature of modern life, even if there is rather less *self*-criticism: and yet progressive success in life is not, as we have seen, exactly widespread.

Our earlier discussions have, however, already revealed that Popper's conception of science, from which his conception of reason springs, is seriously defective. Isn't that right, Scientist? What, you still don't agree? I give up! In any case, for the sake of the argument do please concede the point. Science does not seek truth *as such*, as Popper's methodology in effect presupposes: rather science seeks *valuable truth*. And as we have seen, I believe, recognition of this simple point has profound implications for our understanding of the *progressive* character of science: as our scientific knowledge improves, so too ought our aims to improve; and as our *aims* improve, so too ought our methods to improve as well. Our knowledge about how to improve knowledge improves with the development of science: and it is this, above all, which explains the progressive character of science, something that is missing altogether in Popper's picture of scientific method.

One of the finest successful applications of this more rigorous aim-oriented empiricism to science is to be found in Einstein's development of special and general theories of relativity. At the heart of Einstein's aim-oriented approach to physics there was his passionate desire to discover unity, simplicity, harmony, beauty. And yet this aim, this desire, was, we have agreed, profoundly problematic. We do not *know*, and in a sense we will never know, that Nature has a mathematically unified, harmonious structure. Much less do we know that it has a unified structure of this or that specific type. Thus, the form that our desire, our aim, to discover unity and harmony takes is almost bound to be unrealisable. Our idea as to how Nature is unified – even our vaguest most general ideas – are almost bound to be more or less false. In fact, as Einstein probably alone realised, in about 1902, there was something fundamentally wrong with the whole organising structure of classical physics. For as Planck had in effect shown – even though this was not clearly recognised by Planck himself – classical physics could not conceivably explain how *continuous* light could interact with discrete, *atomic* matter. Light needed to have itself some kind of *particle* character – even though there were overwhelming grounds for holding that light was a continuous wave-like phenomenon. Thus, all organising, unifying ideas implicit in classical physics needed to be thrown into the melting pot. All that Einstein could do was to seek new unifying principles which seemed to resolve implicit clashes in existing ideas. His passionate desire to discover unity, harmony, beauty needed to be shaped, directed, informed by existing knowledge, existing ideas; and these ideas in turn needed to be shaped, developed, moulded by his informed instinctive desire and feeling for unity and harmony. Only by articulating, imaginatively developing and scrutinising new possible unifying aims for physics, in the light of the imperfect knowledge that had already been achieved, could Einstein hope to make fundamental progress. And it was just this process of articulation and exploration of new possible unifying aims for physics that led to the development of the special and general theories of relativity. It was just this which marked Einstein off from aether theorists, who imaginatively developed many different theories of the aether, but who did not get round to questioning whether it was really in the end *desirable* to try to develop an aether theory.

As we have seen, all this has a quite general significance for science. The whole topic of aims is both enormously important and profoundly problematic. Success depends crucially on making a good choice of aim. Indeed the whole success of modern science may be attributed to a good choice of aim. Modern science began with Kepler and Galileo. And what marked off Kepler and Galileo from their contemporaries was not that they alone made experiments or observations: in some respects the Aristotelians were more empirically minded than either Kepler or Galileo. Rather it was their choice of aim: to follow up the idea that, in Galileo's words, "The Book of Nature is written in the language of mathematics" – *simple* mathematics, one might add. This idea, in one form or another, has kept the natural sciences going ever since.

But it is not just to *science* that this Einsteinian procedure of imaginatively articulating, developing and scrutinizing aims is relevant: it is relevant to all that we do, to the whole of life, to all our multifarious individual, social, institutional activities and pursuits. For, quite generally in life, it is both extremely *important* that we make a good choice of aims for the success of our various activities, and often extremely *difficult* to make good choice of

206

aims. Our life aims are all too likely to be *problematic*, either for objective, factual reasons, or for subjective, evaluative reasons, in that their realization does not bring the fulfilment we had hoped for. It is thus all too likely that the aims we pursue, and the aims we believe it to be desirable to pursue are, in reality, ill chosen in that they are more or less unrealizable or more or less undesirable, or some combination of the two. Almost always there will be potential aims more realizable and more desirable than the aims that we are in fact pursuing, and even perhaps think it desirable to pursue. Thus, in our lives, we need to do exactly the kind of thing that Einstein did so successfully in theoretical physics. We need imaginatively to unfurl our desires, our aims, our ideals; and we need to scrutinize and criticize our aims for hidden defects, both of an objective, practical kind, and of a personal, emotional, evaluative kind. We need to inform our objective situation of our feelings, our instinctive desires; and we need to inform our feelings, our instinctive desires, of our objective situation. We need to practise bubble blowing and bubble bursting, as a delightful, playful aspect of life: and as a result, we may be able to get into *life* the kind of progressive success found in science – and to such a marked degree in Einstein's scientific work.

LIBERAL: Bubble blowing? Bubble bursting? What's that?

PHILOSOPHER: Oh, just technical terms. You can perhaps see now, Romantics, why I believe your so-called anti-rationalism is in important respects a *defence* of reason. Unfurling and scrutinizing of aims and desires needs to be put at the heart of rationality, since only by unfurling and scrutinizing possible aims can we do our best to ensure that we have made a *good* choice of aim. Just that which you Romantics have tended traditionally to emphasize – the unfurling and scrutinizing of aims and desires – needs to be put at the heart of reason. This is because only by unfurling and scrutinizing possible and actual aims can we do our best to ensure we have chosen *good* aims. It is supremely important to try to ensure that we have chosen *good* aims, because if we have chosen *bad* aims, then the more rationally, the more intelligently, skilfully, scientifically, effectively, we pursue these *bad* aims, the worse off we shall be, the further we shall be from achieving what we really want to achieve. In short, once we have chosen *bad* aims, reason becomes a *menace*, a hindrance rather than a help. *No ideal for reason can be truly rigorous, truly rational, that does not emphasize the central importance of articulating, developing and scrutinizing aims as an essential aspect of rationality.* All ideals for reason which *fail* to emphasize the central importance of unfurling and scrutinizing aims and desires are irrational and unrigorous just because they are almost bound systematically to lead us astray – granted that in practice we are all too likely to choose bad aims. In my view, this is the essential defect of all traditional conceptions of reason, including Popper's.

Let me now, in just a little more detail, develop the argument that I have just given.

REBELLIOUS ROMANTIC: My God. When I came along here, I had no idea that we would have to submit to an academic lecture.

PHILOSOPHER: (Genuinely contrite) I am sorry. I will make it as brief as I can.

The first and fundamental point to recognize is, of course, the almost inevitably *problematic* character of our aims or desires, whatever we may be doing. There are at least five different ways in which an aim can be misconceived, bad or ill chosen. (1) An aim may be ill chosen because it is unrealisable, either for logical reasons, or for factual reasons, although a slightly modified aim may be both realisable and almost as desirable to attain. Thus, we may seek to discover a method for trisecting a given angle with rule and compass; or we may seek to derive Euclid's fifth postulate, the so-called parallel postulate, from the other axioms and postulates of Euclidian geometry. We may seek to be both loved and feared, an aim which we may well hold to be self contradictory, granted that love excludes fear by definition, as it were. Or again, we might seek to sail West from Africa along the equator to Asia, without coming into contact with land on the way. We may try to develop a perpetual motion machine, a machine which creates an endless supply of energy from nothing. We may seek to go to Heaven, and there may be no Heaven to go to. (2) An aim may be ill chosen in that we pursue it because we judge it to be the best means to the realization of some more distant or more general aim, and this judgement may be wrong: some modified aim may be a better means to the realisation of our more distant or more general aim. (3) An aim may be ill chosen to the extent that there is some slightly modified possible aim which is either more desirable and at least as realisable, or more readily realisable and at least as desirable – or even perhaps both more desirable and more realisable. (4) An aim may be ill chosen in that its realisation brings about all kinds of entirely unforeseen undesirable consequences, although this would not have been the case for some slightly modified possible aim. Thus, we might suppose that it is desirable for everyone to possess their own car, not foreseeing that this will result in horrendous traffic jams, everyone as a result being immobilized. We might decide that it is desirable to promote industrial and economic growth, encouraging everyone to buy and consume goods, not foreseeing undesirable consequences of pollution and depletion of vital finite natural resources. We may decide that it is desirable to pass legislation to ensure that there is a complete equality of wealth, status and power in the community, forgetting that this will have the consequence of placing drastic restrictions on individual liberty. (5) Finally, we may be in a situation of conflict: our chosen aim may represent our best attempt to resolve, or compromise between, our conflicting aims; and in fact some other possible aims constitute a far better resolution of our conflict. (In a sense, both (4) and (5) represent situations of conflict: the difference is that (4) represents unknown or unacknowledged conflict, whereas (5) represents acknowledged but unsatisfactorily resolved conflict.)

Almost inevitably, we may conclude, whatever we may be doing, we will, at some point, have made a bad choice of aim – or *at least* a choice capable of being improved. And as we have seen, the consequences of making a bad choice of aim are serious indeed; for the more *rationally, intelligently, scientifically, effectively* we pursue such a bad aim, the worse off we shall be, the further we shall be from achieving what we really want to achieve. As I have already remarked, once we have made a bad choice of aim, reason becomes a positive *menace*, a hindrance rather than a help, a block to real progress. All our rationalistic methods and strategies – our technology, mathematics, scientific knowledge and so on – designed to help us achieve our aims, in fact work *against* us, and not *for* us. In a situation of badly chosen aims, only the irrational pursuit of our chosen aim, only *irrational* action can enable us to achieve what we really want to achieve, that which is of greatest value. Once we have made a bad choice of aim, irrationality, one might say, becomes a positive rationalistic necessity. As I remarked earlier, absolutely decisive *rationalistic* reasons exist for Romantic suspicion of all concepts of reason which fail to emphasize the central importance of articulating and exploring aims and desires, thus enabling us to make intelligent, wise choices of aims.

All this is serious enough: but there is an even more serious and damaging possibility. We may actually *misrepresent* to ourselves the aim that we are engaged in pursuing. Thus, we may be engaged in pursuing some highly problematic aim A, problematic in one or other of the ways indicated above; it may be possible, however, to resolve these problems by developing a slightly modified, improved aim A* , let us say. Precisely because of the problematic character of A, the fact that it seems to us reprehensible, undesirable, impossible to realise, we may however reject altogether the idea that we are pursuing A (let alone A*) and instead uphold the idea that we are pursuing the quite different, and apparently more desirable, unproblematic aim B. Nevertheless, despite this, we continue, more or less, to pursue A, and continue more or less to recognise as successes those of our actions which take us towards the realization of A, under our cloak of misrepresentation and rationalization.

It is this rationalistic condition which may beset any aim-pursuing activity or enterprise, which I want to call "rationalistic neurosis".

We are, of course, nowadays very familiar with such patterns of confusion through the works of Freud and his followers. The point that I want to emphasize, however, is that "neurosis", as conceived here, is a purely rationalistic or methodological condition, that can beset any aim-pursuing enterprise or aim-pursuing agent whatever:

it is not specifically to be associated with the human psyche or the human mind.

A striking example of an institutionalized aim-seeking enterprise that at present suffers from rationalistic neurosis is, of course, provided by: *Science!* For, as we have seen, the real aim of science is at least to discover humanly *valuable* truth (A), and better still, to help promote human progress (A*). The aim of seeking valuable truth is, however, profoundly problematic. Precisely because of its problematic character, the scientific community as a whole has, in effect, rejected the idea that science seeks *valuable* truth, and replaced it with the idea that science seeks the apparently far less problematic goal of discovering value-neutral truth *as such* (B). As long as the scientific community holds on to the idea that it alone is in a position to make authoritative judgements concerning the intellectual domain of science, the scientific community is more or less obliged to reject the idea that the intellectual aim of science is to seek humanely valuable truth, since the idea that scientists are in a position to make authoritative judgements about what is humanely valuable, for the rest of us, is clearly unacceptable.

Once an aim-pursuing activity or enterprise falls into this condition of rationalistic neurosis, a number of obvious factors will inevitably serve to obstruct our attempts to realise our best interests – which lie in the direction of A or A*. These factors are:

(i) The more intelligent, methodical, honest, thoroughgoing, systematic, or, in a word, *rational* are our attempts to realise our declared goal B, the more unsuccessful, sterile, useless will our actions become when interpreted as attempts to realise our real and best goal A or A*. As a result of misconceiving and misrepresenting our best aim, rationality, once again, becomes a menace, a block to real progress. Only a highly hypocritical, irrational pursuit of our declared goal B can lead to the successful attainment of our real goal A or A*.

(ii) Theoretical investigations into what sort of rationalistic strategies and methods we ought to adopt in order to realise our declared goal B will be profoundly unhelpful, even counterproductive. Not only will we be unable to develop and make explicit the methodological rules we ought to adopt in order to give ourselves the best chances of realising our best interests, A or A*; in addition, any rules that we do develop will tend to be systematically inappropriate for the realisation of A or A*. Explicit methodology, far from being a help, will be a hindrance. The *philosophy* of the enterprise, seeking to lay bare the aims and methods of the enterprise, will be a hindrance rather than a help.

(iii) Even if we are not in practice concerned to pursue our declared goal B in conformity with "official" explicit rationalistic rules and methods, nevertheless our misrepresentation of our basic aim will still be a serious hindrance. For if we pursue our real aim A too openly and brazenly, it will become increasingly difficult to maintain the fiction that it is B that is being sought. In order to maintain the fiction that B is being sought, we will be obliged to pursue A somewhat furtively, token gestures being made in the direction of B. As a result, our pursuit of A or A* cannot be very intelligent or efficient.

(iv) Our genuinely successful actions, our achievements, will present themselves as a series of *problems* to us. For it will be difficult to interpret these successful movements toward A as successful movements towards the different goal B. The more successful we are, the more irrational our actions will seem to be. And the more successful we are in solving these "neurotic" problems —as we may call them – the worse off we shall be from the standpoint of realising A or A*.

(v) We will be unable to formulate clearly to ourselves the real unneurotic problems that need to be solved, if A or A* is to be realised, precisely because this can only be done if our real aims are openly acknowledged. Successful solutions to such problems will seem to arise mysteriously and irrationally, just because they cannot be suggested by explicitly formulated heuristic rules. Discovery, creativity, will come to seem a mysterious, an irrational process. This will further sabotage our attempts to realise A or A*.

(vi) Finally, we will be unable to articulate, and thus solve, the problems associated with A (the cause of the whole trouble) thus adopting explicitly the improved aim A*.

It is, of course, my thesis that science – along with many other of our human social pursuits and activities – suffers from all these rationalistic defects as a result of the misrepresentation of its basic aims, as a result, in other words, of rationalistic neurosis. A striking point to notice in passing is that scientists themselves would be the first to acknowledge features about science which are precisely in accordance with what one would expect to find, if science does indeed suffer from rationalistic neurosis, in view of the above points. Thus, it is generally acknowledged that some of the most strikingly successful scientists seem to violate in their work all the official canons of scientific propriety. Kepler, openly speculative and mystical, is a case in point: and Newton too, in a much more hidden fashion. Faraday was held by his contemporaries to be methodologically naive; and yet again and again it was Faraday who came up with startling discoveries, and not his more scientifically respectable colleagues. Einstein, with his usual accurate and succinct insight, once remarked: look at what scientists do not at what they say they do. Above all, most contemporary scientists are agreed on the complete uselessness of academic philosophy of science (about the only exception to this being that one or two scientists are prepared to acknowledge the importance of Popper's work). Philosophers of science themselves take as their central preoccupation the problem of induction, the problem of understanding how theories can be verified in the light of experience, the problem, in other words, of understanding the success of science. Thus, this central problem shows all the characteristics of a *neurotic* problem (which is, of course, precisely what it is). Furthermore, scientists and philosophers of science are ready to acknowledge that there is something mysterious and extra-rational about scientific discovery, scientific creativity.

Quite generally, in fact, whenever *philosophy* appears to be of little practical help in enabling us more effectively to realise our best aims, we should expect the existence of rationalistic neurosis. The fact that philosophy is quite generally judged to be a somewhat useless activity is perhaps a slight indication of just how widespread neurosis may be. A part of the trouble, of course, is that academic philosophy is itself a profoundly neurotic enterprise; by and large it does not adopt the aim of articulating and exploring

213

actual and possible human social, institutional aims. Academic philosophy *rejects* this commonsense conception of philosophy.

The consequences of rationally pursuing ill chosen aims are, as we have seen, serious enough. The consequences of *misrepresenting* aims, of allowing an enterprise to fall into rationalistic neurosis, are however in some ways even more serious. For if we simply pursue bad aims, that are however clearly acknowledged, it is fairly easy at least to discover and to see what is going wrong; in order to put things right, new aims need to be adopted. Once an activity has become rationalistically neurotic it is however much more difficult to discover what has gone wrong, and what needs to be done to put things right. It is an essential feature of rationalistic neurosis that our very efforts to put things right may actually make things worse. As a result of the misrepresentation of aims, attempts to ensure that the activity or enterprise pursues its declared aims more conscientiously and more honestly actually serve to obstruct real progress. And the demonstrably counterproductive character of attempts to clarify the aims and methods of the enterprise serve to cast doubts on the practical value of seeking to clarify and enhance our understanding of aims and methods. Success will be achieved *despite* such work, not because of it. In short, rationalistic neurosis, once established, tends to maintain itself. It contains within itself its own mechanisms of self-preservation. Indeed just these mechanisms constitute the greatest hurdle to the effective practical application of the ideas that I am putting forward here. Whatever else it may be, my suggestion is in a sense a contribution to *philosophy*. Who has ever heard of philosophical ideas, solutions to philosophical problems, having useful and valuable *practical* applications? What scientist is likely to think that a contribution to the philosophy of science has any relevance to science itself?

What is to be done? Well, it is very easy to see the kind of strategies that it is in general *desirable* to adopt, whatever we may be doing, whether we are pursuing intellectual or non-intellectual goals. It is easy to see the kind of strategies we can adopt in order to give ourselves the best hope of ensuring that reason works *for* us, and not *against* us – so that we act in a genuinely rational fashion. *First*, of course, we need to make every attempt to ensure that we have accurately represented to ourselves the aims that we are, in fact, pursuing – in order to avoid rationalistic neurosis. *Second*, given that we are pursuing some aim A, we can develop more restrictive, precisely defined versions of A, so that as we draw closer to the attainment of A, we can have precise versions of A to choose from. At the same time we can develop looser, broader, more loosely defined versions of A, so that if A turns out to be unrealizable, we are not simply at a loss. *Third*, we can subject our chosen aim A to a barrage of imaginative exploration, development and criticism, in an attempt to develop A in more desirable or more realizable directions. Thus, on the one hand we can look at A with an upsurge of optimism and indulgence, as it were, seeking to develop more desirable, ambitious, valuable versions of A; and at the same time we can look at A in a highly critical, harsh, pessimistic spirit, probing A for hidden snags and difficulties, searching for possible obstacles to the realization of A, or undesirable consequences hidden in A, thus developing more practical, realizable versions of A. It is, of course, always possible that this process might lead to the discovery of an aim which is *both* more desirable *and* more realizable than A. *Fourth*, we need to articulate and expose conflicts in our aims and desires, in an attempt to develop better resolutions of such conflicts. *Fifth*, we need to ask *why* we are pursuing our chosen aim, both in the rationalistic sense of what more distant or more general aim we hope to realise, and in the *historical* sense of what first *caused* us, or prompted us, to pursue our aim. Almost all our aims evolve from previously pursued aims: looking at historical origins enables us to discover whether we have retained and developed what is best from the past, and at the same time have given up what has become inappropriate to the changed circumstances of the present. All this is, of course, especially relevant for the detection of rationalistic neurosis. *Sixth*, we need to be constantly alert to the possibility of rationalistic neurosis: whenever the very process of articulating and exploring aims, and developing methods designed to help us realise our best choice of aims, appears to be curiously unhelpful, and even counter-productive, we should at once suspect the existence of rationalistic neurosis. Far from concluding, in these circumstances, that the philosophy of our activity or enterprise is inherently useless, we should conclude that there are serious faults with the philosophy that actually influences our actions, in that this philosophy *misrepresents* the aims that we are, in fact, pursuing. It is just where philosophy appears to be most useless that it has, potentially, the most dramatically helpful contributions to make. *Seventh*, and finally, we need to choose and develop those *aim-dependent* methods, rules, strategies, techniques which are most appropriate to, which give us the best hope of realizing, our best choice of aim or desire. Aims and methods need to evolve in harmony with one another – as in the case of Einsteinian physics.

All these rational strategies are just as relevant and desirable on the level of individual, personal action, as they are on the level of inter-personal, institutional, social aim-pursuing. They have, in fact, a completely general, universal application. They are relevant whoever, or whatever, may be doing the aim-pursuing, whether individual person, group of people, institution, animal, robot, or whatever.

You, Scientist, I am sure, will recognize some of these strategies from our earlier discussions about science, especially the *fifth* strategy.

Our whole discussion, so far, might be encapsulated in the remark: *An essential requirement for rationality is that we choose our aims rationally.* This point seems to me to be of such fundamental importance that it deserves perhaps to be called: *The Fundamental Theorem of Rationality.* The proof of the theorem can be presented as a *reductio ad absurdum* argument!

Let us suppose that reason has nothing to do with choosing aims, but merely provides rules, methods, strategies, techniques, which help us to realise aims, chosen extra-rationally. In this case, if we make a bad choice of aim, as we are all too likely to do, for quite general reasons, then the more rationally – and so effectively – we pursue this bad aim, the worse off we shall be. Our assumption that reason has nothing to do with helping us to choose good aims has led directly to the conclusion that "reason" must tend systematically to work against our best interests, failing to help us to achieve effectively that which is of greatest value. Hence our assumption must be wrong. It must be the case that genuine rationality does involve the potential choosing of aims. Q.E.D.

To choose aims rationally is *not* of course to allow *Reason* to determine for us what our aims ought to be. On the contrary, the proper rational function of reason is precisely to help *us* to make the choices, the decisions that we really do want to make. Thus choosing aims rationally is to choose aims as a result of having provided ourselves with a rich store of possible aims both accurately examined and accurately experienced in imagination, so that our capacity to make a good choice, a wise choice, the choice that we really do want to make is enhanced.

Reason is *desirable* because it helps us to discover and to achieve that which is genuinely desirable. Thus if we are to be rational in our approach to Reason we need to consider carefully whether the concept of Reason that we accept is the most *desirable* that is conceivable, or whether perhaps our concept can be made a little more desirable. If Reason does not *feel* good; if it does not seem to be genuinely helping us to discover and realise our most desirable objectives, then something serious is wrong. Either we are *using* Reason improperly; or there is a rationalistic defect in our very concept of Reason (or, of course, we have lost all freedom of action, and Reason has become useless). If we can find some reason why our concept of Reason, appropriately used, should have a tendency systematically to lead us astray, failing to help us get what we really want to get, then *ipso facto* there is something wrong with our concept of Reason. Our very concept of Reason needs to be assessed and developed by taking into account our instinctive feelings concerning its value. In short, aim-oriented rationalism needs to be applied to itself.

Rational choosing of aims might be imagined in the following terms. We use our mind, our imagination, as a kind of display screen upon which we project as accurate and honest a representation of the aims we are pursuing as possible. (On the institutional, social level we represent institutional, social aims as accurately and honestly as possible within public means of discussion and communication – artistic/intellectual communication). We make every attempt to ensure that we have accurately represented the aims that we are in fact engaged in pursuing. We ask: In our heart of hearts, is this *really* what we want to achieve in pursuing this activity? Do our actions make sense given the supposition that this is what we are seeking? Is there perhaps some aspect of our aim which we are systematically ignoring or distorting?

We then seek to develop this represented aim in more desirable, realisable directions. With our intellects we look for hidden objective defects in the aim: we try to develop more easily realisable aims. And, at the same time, with the more imaginative aspect of our mind, accurately sensitive to our feelings and desires, we develop more desirable possible aims, constantly seeking to assess by means of accurate vivid imagined experiences, whether realisation of the aim in question would really have the value, the desirability, that we might suppose. These two processes of imaginative intellectual development and assessment of possible aims or desires proceed harmoniously inter-related, precisely so that we can give ourselves the best chances of discovering that aim which is both most desirable and most realisable in our circumstances. The unit of attention is the *aim*: Reason helps us to improve and assess this from *both* standpoints: realisability and desirability; factually and evaluatively; objectively and emotionally. Genuine rationality actually *requires* that these two kinds of consideration work in conjunction with each other. Reason, I hope it is clear, helps us to discover and to decide in what directions we want to develop our aims, by providing us with a rich store of accurately imagined, critically examined possible aims, thus enhancing *our* capacity to choose wisely and well.

We then modify our actual aim in the directions which seem to be desirable, and develop appropriate new methods and techniques designed to help us realise this new modified aim, old methods and techniques that have become inappropriate being allowed to lapse.

From the vantage point of aim-oriented rationalism, we can see quite clearly what are the rationalistic defects of Popper's critical conception of reason. To begin with, of course, Popper's conception of reason suffers from the defect of being applicable in the first instance to *intellectual* entities – theories, ideas, proposals – and only indirectly to desires, aims, actions. But even if we ignore this, there is still a serious inadequacy in what Popper has to say. It is, of course, important if we are to meet with progressive success in our actions, that we can distinguish real success from false success – from hallucinations of success, if you like. The capacity to recognise failure is, of course, *necessary*: but on its own it is almost completely useless. If we have no capacity to generate *good* possible trials, *relevant* possible actions, our chances of coming up with a successful trial or action will be infinitely remote. We will spend our time rejecting failure after failure, with splendid Popperian zeal, getting nowhere. Only if we have a reasonable capacity to come up with good candidates for success are we likely actually to achieve success. In short, only if our *aim* is reasonably good can we hope to achieve success. And of course in practice we always do have a more or less good aim, a more or less good idea of what it is that we seek to achieve, which guides us to the generation of trial actions. And our aims are always more specific than the archetypal Popperian aim of seeking to discover factual truth, even in science. How good our choice of aim is will thus crucially affect how good our trial action is and thus how successful these actions will be. A rigidly maintained bad aim will permanently sabotage success, despite rigorous observance of Popper's critical methodology. Hence the supreme importance of doing everything we can to ensure that our choice of aim represents a good choice, our aim being delicately and progressively developed in the light of the success and failure of our actions, successful action leading to the progressive improvement of our aim, and hence to the progressive improvement of our capacity to generate further successful action.

Our science, our culture, our social institutions, need to be sensitively arranged around human action, so as to encourage and promote rational human action in this sense – so as to encourage and promote action to evolve progressively in humanly valuable directions.

There is now one important point that needs to be made. The whole purpose of rational exploration of aims is to help us to improve and develop the aims that we are actually pursuing. What matters, in the end, is that our actual aims evolve *as if* we had rationally explored possible aims, and put into practice our genuine discoveries. Rational *action* is in other words more important than rational cogitation (except of course that cogitation can itself be a kind of action, pursued for its own sake). The more instinctively, effortlessly, easily, enjoyably, we are able to develop our aims in fruitful, realisable directions, without elaborate conscious exploration of possibilities, so the more successful we will be. It is desirable, in short, for aim-oriented rationalism instinctively to inform our actions.

The great danger of a too elaborate, self conscious rational, imaginative unfurling of aims or ideals is that a vast gulf may begin to grow between genuinely *good* possible aims developed by rational exploration and *actual* aims. Just because of the existence of an immense discrepancy between what we have discovered it to be in principle desirable and realisable to pursue, and what we are in fact engaged in pursuing, we may find it impossible to develop our actual aims in the directions which we desire. Aim seeking activities, both individual, and even more institutional and social, tend to have a certain resistance to change. The discrepancy between reality and the ideal may lead us to despair, or to hate reality for falling short of the ideal. The very *success* of the process of rational exploration of aims leads to the *sabotaging* of rational action, and not to its advancement. (Once again, reason has been found to be counterproductive if not delicately handled.) If aim-oriented rationalism is to help lead to the flowering of our *lives* then it is important that we use rational exploration of aims to help us to discover and to develop that which is of greatest value which is implicit in our actual aims, actions, life. Thought and imagination need to be used intelligently to *enhance* freedom, not to undermine freedom.

The great thing, in other words, is not to press too hard. For if one does, life becomes a misery and all meaning and enjoyment in one's life will drain away. If aim-oriented rationalism is to work *for* us, then we need to use it to point our noses first in the direction of *enjoyment* of life, then perhaps in the direction of improving the enjoyment in various ways – sharing it with others, taking others into account, and so on (whatever our views may be in this direction.) Once aim-oriented rationalism has become *enjoyable* then it can more or less look after itself, and one can attend wholly to more important matters. (It is in any case always a bad mistake to be too concerned with questions of *method*, with means rather than ends: aim-oriented rationalism itself tells us this!)

There is, of course, another possibility not yet considered. Rational exploration of an aim actually being pursued may reveal that the aim cannot be developed into fruitful directions. In this case, we will need to develop strategies which help us to *discard* the associated aim-seeking activities, so that we can come to dissociate ourselves from the unwanted aim, treating it like an unwanted piece of clothing. Aim-oriented rationalism helps us to discover the desirable balance between developing aims and desires, and abandoning unwanted aims and desires.

All this, in effect, provides us with a rational method for articulating, developing and assessing philosophies of life. For the aims and methods we in fact pursue and practice constitute our actual philosophy of life; the aims and methods we believe we in fact pursue, desire to pursue, and hold to be ideal to pursue, constitute our conscious philosophy. Aim-oriented rationalism provides us with a rational method for helping us to develop our philosophies of life in the direction in which we really do want to develop them. Instead of a philosophy, a way of life, choosing us, we can choose it, thus enhancing our freedom.

The implication of our argument is that, at the centre of intellectual, rational concern, we should put philosophies of life – or philosophies of various aspects of life. Our best science, technology, mathematics, objective intellectual enquiry, develop the more objective, public, inter-personal aspects of our philosophies of life. Our best literature, drama, poetry, music, art, explore and express the more emotional, personal, intimate, instinctive aspects of our philosophies of life. If we are to develop an aim-oriented culture, a culture which really does put aims, philosophies of life, at the centre, then we need to enhance the artistic aspects of science, and the rational aspects of literature, so that we can discuss our aims playfully, imaginatively, and in an emotionally honest and pleasing way, and at the same time objectively, thoughtfully, critically – so that the discussion is both delightful and useful, pleasing and practical. In this way, we may gradually discover how to share our ideas, our discoveries, our problems, our experiences, our philosophies, easily, delightfully, imaginatively and *usefully*. At present there is either a tendency for people to try to bully others into accepting their own ideas as the Truth; or there is a tendency for people to keep quiet about the whole subject, in reaction to all the bullying, the dogmatism, the thundering. Both attitudes are unproductive. The possibility of an enjoyable sharing of ideas between friends disappears. And it is just this excluded possibility which would undoubtedly lead to the most rapid and valuable learning and mutual understanding between people. Differences between people could become a delight, an opportunity for discovery, rather than a threat.

What is being proposed here might be regarded as an extension of conventional empiricism – an extension of aim-oriented empiricism. For aim-oriented rationalism provides a framework for testing philosophies of life, ideals, values, against our personal experiences, our problems, frustrations, joys, difficulties, successes. In science, "experience" is required to be severely de-emotionalised, depersonalised, intellectualised, as it were – a small segment of *personal* experience. Our argument, in indicating the rational need, or desirability, of putting philosophies of life at the centre of intellectual attention, and in calling attention to the role of personal feelings, desires, and actions in developing and appraising such philosophies, in fact emphasizes the centrality of the whole, rounded, integrated personal experience. It is against this that we ought primarily to think, if we are to think in a truly rational fashion, and not primarily against the depersonalized, de-emotionalized, fragmented experience of science.

I have described aim-oriented rationalism as if it were a very individualistic activity. But, of course, it would be foolish (as well as impossible) to ignore the ideas, the discoveries, the successes and failures of others even if our aims were defined in a wholly individualistic fashion. The best aims of life have, however, a cooperative, inter-personal or social character. In pursuing such aims rationally, aim-oriented rationalism becomes something to be practised by many people, cooperatively.

There! That is what I mean by *Reason*! A rational life, and a rational society, is simply a life, a society, that takes seriously and light-heartedly this game of bubble blowing and bubble bursting. Do you see, Liberal, why I believe aim-oriented rationalism can heal the undesirable gulf that has grown up in our world between the "public, objective" world, and our "private, personal, subjective" worlds? There is in existence today an immense amount of objective, technical skill and knowledge, which enables us to realise all kinds of goals that would have been almost inconceivable in earlier times. And there is in our society a great deal of what might be called humane knowledge – sensitivity and understanding, intelligence and insight, concerning the feelings, problems, motivations, desires of individual people. But what is so lacking is the capacity to bring together, to combine, these two kinds of knowledge and understanding. As a result, science, technology, specialist academic learning, tend to proliferate in ways which ignore the personal, the emotional, the needs and problems of people in their lives. Commerce and industry develop in directions which often seem to be of questionable value to each of us individually. In fields such as education, psychiatry and medicine, where the two kinds of knowledge and understanding are clearly both needed, one often finds conflict and breakdown of communication between upholders of romantic and rationalist attitudes and values, rather than an easy synthesis of these attitudes. The same kind of gulf exists perhaps most disastrously in the realm of political action and thought. Our whole modern world pursues aims which we know must, in the long run, for our great-grandchildren, prove disastrous.

Yet again, as a result of this gulf between the objective and the personal, emotional personal problems tend to be viewed, not as problems around which all our thinking and planning should arrange themselves, but rather as a kind of *illness*, an incapacity, which needs *treatment*. Precisely the objective, social intellectual or philosophical aspects of personal, emotional problems are removed from view. The very task of speaking out clearly, simply and directly about how aspects of the modern world create personal problems for one becomes almost impossible since the mere confession that one does have serious emotional problems means that one is ill, incapacitated, and hence in no state to make objective judgements.

All this has I believe arisen because of the failure to put aim-oriented rationalism into practice. Aim-oriented rationalism provides a framework for healing the gulf that exists in our society and culture between mind and heart, thought and feeling, the objective and the subjective, the impersonal and the personal, fact and value, science and art, idea and desire. Our argument has demonstrated the rational need for this, and just how desirable it is. A *rational* society would be a society in which aim-oriented rationalism would be commonplace, culture and social institutions being arranged on aim-oriented rationalistic lines, culture and education being designed to stimulate, encourage and promote instinctive, easy aim-oriented rationalistic action. Lack of wisdom in our times – insofar as there is a lack – in people, in our various cooperative, social, political pursuits and enterprises is, I conjecture, a direct consequence of our failure to practice aim-oriented rationalism. The crisis of our times is due to the evolution of our social, environmental, cultural circumstances unguided by practical, active aim-oriented rationalism.

RATIONALIST: I have never heard such a rigmarole of rubbish. For a start, you haven't even defined your terms. What do you mean by Reason, by Rationality?

PHILOSOPHER: (laughs) Oh, you want to play the disreputable game of definitions, do you?

Deliberately, I leave the meaning of Reason or Rationality as vague as possible. In using these words I assume only that there is some loosely defined set of general strategies, principles, procedures, methods, rules, skills, techniques, which serve to help us realise our aims, overcome our difficulties, solve our problems, whatever precisely we may be doing. We act "rationally" insofar as we act in accordance with these strategies, and "irrationally" insofar as our actions blatantly violate these strategies. Thus, in various contexts, "rational" as used here is more or less equivalent to intelligent, skilful, wise, knowledgeable, clever, practical, perceptive, brilliant, creative, imaginative, scientific, methodical.

The importance of aim-oriented rationality is that it supplies a completely general strategy for the desirable, rational use of all aim-dependent strategies. In emphasizing the central importance of choosing aims rationally, and choosing methods appropriate to our chosen aims, aim-oriented rationalism helps us to select those *other* strategies which genuinely help us to achieve what we really want to achieve.

LIBERAL: Shouldn't one just say that to be rational is to be *logical*? Is not the essential requirement for rationality that one's theories or beliefs be logically consistent?

PHILOSOPHER : Not at all. It would be profoundly irrational to demand logical consistency at all times. All that the discovery of inconsistency reveals is that there is something wrong somewhere – which we should have guessed anyway. It may be very difficult to discover what exactly is wrong, and how our ideas can be improved. In the meantime we can live with the existence of the inconsistency quite rationally – acknowledging the existence of the *problem*. Just think what a disaster for science demanding consistency would be. We would have to throw away either general relativity or quantum theory, since the two are incompatible. Never be bullied by pseudo-Rationalists into manufacturing instant consistency!

ROMANTIC: It seems to me that this aim-oriented rationalist stuff is just plain common sense.

PHILOSOPHER: It is!

ROMANTIC: Then why do you make such heavy weather of it? Why make it into such a big issue?

PHILOSOPHER: For three reasons. First, it is, I believe, for a number of reasons, really quite difficult to put aim-oriented rationalism into practice. All kinds of factors discourage us. We are inclined to think that the one thing that we can know for certain in an uncertain world is what we *want*. Challenging basic aims and ideals tends to be experienced as a threat to our very identity, in that our sense of identity is bound up in our aims. If we have actively committed ourselves to trying to realise some aim, it is not easy to acknowledge that the aim might have been misconceived all along. It is very difficult to combine practical commitment with theoretical tentativeness indicated as desirable by aim-oriented rationalism. We tend to be defensive concerning our basic aims and ideals. Instead of *prizing* diversity, we feel threatened by it.

In view of all these factors tending to discourage us from putting aim-oriented rationalism into practice, one is inclined to say that those pursuits which pride themselves on their *rationality* have a supreme duty to practise and propagate aim-oriented rationalism. What do we find? Exactly the opposite! It is the scientists, the mathematicians, the academics, who are the worst offenders! Our very ideals of reason, of scientific, rigorous, exact, rational thinking *suppress* aim-oriented rationalism instead of promoting it. Science, education, academic learning, standards of rationality and rigour, have the effect, in some ways, of depriving people of their common sense, their capacity to think straight, to act rationally. This is my second reason for emphasizing the whole argument. And my third reason arises from the fact that I believe that the *consequences* of irrational scientific, academic and technological thinking – lacking basic elements of common sense – have been serious indeed for our modern world, for our whole way of life.

SCIENTIST: The point I think I find the most unconvincing is your assertion that you have provided us with a purely rationalistic, methodological reinterpretation of certain key ideas of Freud – such as neurosis, repression, unconscious desire, rationalization. It seems to me that your account here was really rather psychologistic in character.

PHILOSOPHER: A few remarks, will, I believe, make it quite clear that the whole thing is entirely rationalistic and methodological in character. But the remarks will be a bit abstract, I'm afraid. The others may be bored.

226

REBELLIOUS ROMANTIC: Whatever you do, don't encourage him any more.

SCIENTIST: But I really do *want* an answer to my objection. Reason is supposed to help us achieve what we *want*, is it not?

PHILOSOPHER: Very well. Let us suppose we have some aim pursuing entity, E, which need not at all be conscious, or have a mind, but which possesses the following characteristics:-

(i) E has an imperfect "knowledge" of its environment and itself, and is not "aware" of all the implications of the knowledge that it does possess. In speaking of "knowledge" and "awareness" here, we are not necessarily presupposing that E has a mind or is conscious : we are referring only to information stored in E, beliefs "programmed" into E, as it were, capable of influencing E's choice of aims and E's actions.

(ii) Some of E's aims can be ordered by E in terms of "preference" or "desirability".

(iii) Some of E's aims are hierarchically ordered in that certain aims are pursued in order to attain more distant or general aims.

(iv) E can represent to itself aims that it may be pursuing in some symbolic form, such as in terms of a picture, a mobile dot on a chart, or in terms of a language. In addition, there is some kind of feedback mechanism between actions performed by E, and goals that E represents to itself as being pursued. A too blatant discrepancy between actions and represented aims will lead to a modification either of action, or of represented aims.

That is all that is required for all the rationalistic, methodological considerations that I have been discussing to be applicable to E. In particular, it is enough for E to be capable of falling into "rationalistic neurosis", as defined above. E may pursue some aim, A, and represent to itself that it is pursuing a rather different aim, B. In fact if E does not adopt the strategies of aim-oriented rationalism in its aim seeking activities, then it is almost inevitable that E will fall into rationalistic neurosis. I have not assumed, however, that E is a person. E *could* be a person: but equally E could be a group of people, or an aim-seeking activity or enterprise that can be conceived of as preserving its identity throughout a gradual change of personnel, such as a social institution, a business, a school, a newspaper, a political or religious movement, a philosophy of life, an art form, a culture, a society, a way of life, or an academic discipline such as mathematics, theoretical physics or literary criticism; equally, E might be an animal, or a sufficiently sophisticated robot.

The reinterpretation of Freud that I have proposed not only utterly transforms the intellectual standing of Freudian ideas: it also enormously extends the range of application of Freudian ideas, as these few remarks indicate.

SCIENTIST: Earlier you said that your new ideal for reason led to a completely new conception of the social sciences, which at last put the social sciences on a truly rigorous footing. What did you mean by that?

REBELLIOUS ROMANTIC: Oh, no!

PHILOSOPHER: It's very simple: the primary task of the social sciences should be to help us to work out how we might pursue our various personal, institutional, social, political aims and activities in a rather more aim-oriented rationalistic fashion, in a way which seems to be genuinely more desirable to the people involved.

Traditionally, the social sciences have of course been conceived of, on analogy with the natural sciences, in very different terms. The aim of the natural sciences is to improve our knowledge and understanding of natural phenomena; analogously, it is assumed, the aim of the social sciences is to improve our knowledge and understanding of social phenomena. Improved knowledge of natural phenomena leads to valuable new technology; likewise, improved knowledge of social phenomena may lead to valuable new social technology.

My argument has, however, provided us with a quite different and, in my view, far more valuable and rational conception of the aims and methods of the social sciences. The aim of the social sciences should not be to apply the methods of the natural sciences to the task of improving our *knowledge* of social phenomena: rather the aim of the social sciences should be to work out how we

228

may apply the methods of the natural sciences – so amazingly successful within the restricted field of science – *to helping us to realise our various personal, social aims and desires*. The aim of the social sciences should be to work out how we can employ, in a humanly *desirable* way, a suitable generalization of scientific method in all our various human, social, institutional, political activities and enterprises, so that we can get into all these pursuits some of the progressive success, on a *human* level, that is found so strikingly on an intellectual level, within the natural sciences. The social sciences become, in other words, *methodologies* of personal, social, institutional pursuits. What a methodologist is to physics, let us say, so an economist is to economic pursuits in general, a psychologist to personal pursuits; and so on. Or, to put the same point the other way round: our present discussions amount to contributions to the *sociology* of science (in the new sense of sociology). What I am trying to do here for science, other social scientists should seek to do for industry, schools, the media, government, international relationships, personal relationships, developing understanding between people, and so on. Up till now, it has not really been possible to take seriously this suggestion just because scientific method has seemed almost exclusively restricted to attaining the aim of acquiring knowledge of factual truth. Scientific method has thus not seemed to be desirably or usefully applicable to the realization of other personal or social aims – such as happiness, love, justice, mutual understanding between people, and so on.

Our development of a more *rigorous* conception of scientific method has, however, changed all this. The more rigorous conception of scientific method of *aim-oriented empiricism* – which so successfully explains the progressively successful character of science – can be generalized, as we have seen, to become *aim-oriented rationalism*, a methodology of universal value, whatever we may be doing, whatever our aims and values may be, whatever our philosophy of life may be. Thus we have the *desirable* possibility: the aim of the social sciences is to work out desirable ways of pursuing our various personal, social enterprises in a rather more aim-oriented rationalistic fashion, the human, institutionalized enterprise of science simply being a special case of this.

SCIENTIST: I am not sure how social scientists would react to being described as social methodologists. One last question: what did you mean when you claimed that aim-oriented rationalism is a general methodology of problem solving and also, at the same time, a theory of rational creativity?

PHILOSOPHER: The thing can be put like this. Traditionally, rationality and creativity tend to be regarded as being somewhat distinct. People who are extremely clever, skilful, intelligent, knowledgeable in their field, are not always held to be correspondingly *creative*, in that they have an especially striking capacity to discover, or make, new things of value. Creative people, on the other hand, although possessing sufficient skills, techniques, intelligence in order to practise their craft successfully, often do not seem to be noticeably more skilful, intelligent, etc., than their less creative colleagues. With respect to conventional conceptions of reason, creativity comes out as a mysterious, personal extra-rational plus. Aim-oriented rationalism, however, completely changes this situation.

Quite generally, a problem may be characterised as an aim A, a provisional idea as to a route, R, to the realisation of A, and a block B, to the realisation of A. (If we had no provisional idea for a route to our aim, not even on analogy with some previously solved problem, then we would scarcely be able even to conceive of the aim A in the first place.) A problem is thus a triad [A,R,B]. The first rule of rational problem solving is perhaps: try out whatever occurs to you; and if that doesn't work, try to get a clearer understanding of the problem to be solved. We need to articulate both R and B, so that we can work out how to develop means for overcoming B, or for developing some alternative route round B. But we also need to articulate and scrutinise A. We may find that there is some slightly different aim A* which is actually more desirable to attain than A itself, thus perhaps profoundly changing the nature of our problem. The new more desirable aim A* may even be easily realised. The solution to our original problem may actually involve *changing* the problem we began with. As a result of articulating, imaginatively developing and scrutinising aims, thus putting ourselves in a position to choose our aims rationally, we put ourselves in the position of being able rationally to develop and choose our problems. Aim-oriented rationalism helps us to ensure that the problems that we choose really are the problems that we need and wish to solve – and may be able to solve.

The hallmark of creativity is, I suggest, to have the capacity to nose out, and realise, aims that are especially valuable to realise. The creative person is the person who chooses and develops problems whose resolutions represent things of real value. It is just this, however, as we have seen, which aim-oriented rationality helps to make a progressive rational process – whereas of course traditional concepts of reason and traditional heuristic rules of problem solving, though providing techniques which help us to solve given problems, provide no techniques which help us to choose and develop problems in fruitful, valuable directions. Creativity and discovery only appear as extra-rational because we cling to unrigorous, non aim-oriented conceptions of reason. There is reason to suppose that the institutionalisation and socialisation of aim-oriented rationality would render creativity a much more public, inter-personal, communicable, learnable quality, creative activity being far more something which can profitably go on amongst many people, rather than being the exclusive and non-communicable activity of individuals.

REBELLIOUS ROMANTIC: Right. That's it. Reason now claims to take over *everything*, even creativity. I have had more than enough.

PHILOSOPHER: Just one last small point, please, and then I promise I will stop.

REBELLIOUS ROMANTIC: No!

THE REST (Reluctantly): Let him have his say.

PHILOSOPHER: All that I wanted to say is that I am inclined to believe that the ideal state of affairs is a relationship of harmony between ourselves, and between ourselves and Nature, so that our instinctive desires, our spontaneous feelings and impulses to act, are precisely appropriate to the realisation of that which is of value.

231

Something of this ideal relationship is perhaps to be found in Nature itself. The instinctive impulses of animals, their responses of fear, rage, desire and gratification, are delicately adjusted to their circumstances, to their objective problems, so that instinctive reactions are appropriate and, by and large, successful. All this is, of course, contrived by natural selection. Just how delicately contrived this harmony of instinct and environment can be is perhaps indicated by the fact that many animals cease to breed, or cease to care for their young when kept in captivity, wrenched from their natural habitat.

Something of the same harmony at the level of human life, is, perhaps, to be found amongst the Pygmies described by Turnbull. In his book, Turnbull suggests, convincingly I think, that the striking differences to be found between the Pygmy way of life, and the way of life of the villagers, is due to the fact that, whereas the Pygmies are hunters and gatherers, in a sense being provided for by Nature, the villagers depend on agriculture, which requires long term planning, methodical clearing of land, the forest thus becoming a constant encroaching threat to their livelihood. Whereas the Pygmies can lead an instinctive, day to day existence, the villagers cannot: technological advancement has to some extent disrupted the relationship of trust and love, turned Nature into a threat and a danger. And as our social and technological arrangements have evolved, and have become ever more sophisticated and complex, the easy, informal give and take between circumstances, action, feeling and desire has been increasingly disrupted. Our instinctive emotional responses become increasingly inappropriate and unhelpful; so, as far as our public, objective, technical, institutional activities are concerned, we try to ignore them. As a result, the gulf grows wider still: the public, objective aspect of things develops without our instinctive emotional responses; our instinctive, emotional responses become increasingly inappropriate to our new environment. Emotionally we can no longer understand our world. We have failed to educate our emotions, our instinctive desires; and our objective public world becomes increasingly ill-informed, uneducated from the standpoint of our feelings. We divide problems up into two categories: objective and subjective, public and private, social and emotional, intellectual and personal. We lose the capacity to design our cities so that they please our feelings. Our social order develops in ways which disregard our personal frustrations and aspirations. We no longer know how to move intelligently and effectively from felt frustration to objective problem, and from objective problem to felt frustration. Thinking becomes increasingly technical and complicated, just because it has lost sight of the capacity to present technical ideas in a simple and instinctively acceptable form. Science loses sight of its human point; it loses its human heart, and eventually loses its mind, in that arguments, in order to be considered at all, have to satisfy certain strict requirements, the legitimacy of which can no longer be effectively criticised or discussed within the scientific context. (Once standard empiricism is accepted, essentially only *testable* ideas are permitted to enter into the scientific domain: standard empiricism, itself being an untestable idea, thus banishes itself from the realm of the scientifically discussable, and thus from effective criticism.) Our very standards of rationality demand that "subjective" feelings, impulses, desires, be ignored.

Aim-oriented rationality is a delicately contrived methodology designed to help heal this gulf between the objective and the personal, the factual and the emotional, thought and desire. (It was just the harmony between instinctive impulse, feeling and desire, and objective thought, that was such a striking feature of Einstein's way of doing theoretical physics.) I have tried to show that there are invincible rationalistic grounds for taking our desires, our feelings, our impulses into account if we are to act and to think in a truly rational fashion. And I have tried to indicate that it is *desirable* to allow our desires, our feelings, our impulses to be unravelled by our minds, delicately explored by "objective" modes of thought, for emotionally unforeseen snags and difficulties, so that what feels right can become a technical, objective, possibility. In order to act rationally – that is in a way which enables us to achieve, to create, that which our heart desires – we need to develop the capacity to use our minds, our imagination, as a display board upon which we can project our desires, our emotional needs and problems, so that we can at one and the same time explore with our objective intellects possible snags and difficulties and at the same time accurately anticipate with our feelings, our hearts, what it would be like to *experience* what we envisage, the state of affairs we aim to realise. Our mind informs our heart; and our heart informs our mind. In this way we can perhaps hit upon that which is both objectively feasible and emotionally desirable. Our objective world can learn of, and become intelligently responsive to, our feelings: and our feelings, our instinctive impulses can become intelligently appropriate to our objective world. Gradually we may be able to create a world about which we can begin to feel good.

CHAPTER TEN

A SONG TO HELP MAKE THE WORLD HAPPY

SCIENTIST: (To the Philosopher) At last, then, we have your complete argument before us.

PHILOSOPHER: No. Not yet.

SCIENTIST: *What?*

PHILOSOPHER: We have almost arrived. But the most important step of all still remains to be taken. Once we agree that —

REBELLIOUS ROMANTIC: We agreed, I thought, that your last speech really was to be your last.

PHILOSOPHER: Please! Do please let me just outline this last step. It is so important. Especially for us non-specialists.

(A moment's silence is quite enough for our Philosopher to take as a deafening call to continue.)

It can be put like this. If science is to proceed in a truly rigorous, objective and rational fashion, then science cannot conceivably be left in the hands of the expert, professional scientists. Quietly and firmly *we* must take possession of science, and actively concern ourselves with it, and engage ourselves with it, as an aspect of our relationships with each other and with the world. Science is not the exclusive property of experts: it is *ours*; it is a community affair, something which goes on between people as an aspect of life. Science is something that *we* do, *we* create, *we* share. It is a part of our attempts to create better relationships between ourselves, and between ourselves and Nature – relationships that are less painful, restricting, frustrating, embittered; more knowledgeable, understanding, appreciative, sensitive, delightful, compassionate. Science is our expression of all this. It is our shared discoveries, perceptions, explorations. It is our ideas, proposals, suggestions, designed to help improve things. Science is not something done to us, for us, or at us, by experts; *we* do it. Science is the chorus of our songs sung "in order to help make the world happy" as the Pygmies would say. Only if *we* take possession of science in this kind of way can we hope to have a science that is both humanly desirable and truly rigorous, objective and rational.

SCIENTIST: It's the last point that I don't see. A case can I think be made out for saying that it would be *desirable* for non-scientists to take a more active interest in science: but why should this amount to an enhancement of rigour? When one takes into account the general level of discussion of issues in society today, the popularity of all kinds of absurd myths and fairy tales, one can only conclude, surely, that the kind of popular take-over of science that you seem to be proposing could only result in the *destruction* of science, the sabotaging of scientific rigour and objectivity.

PHILOSOPHER: Not for a moment am I proposing that experts should simply be brushed aside, ignored. My point is that it is absolutely essential for experts and non-experts to *share* science, and above all communicate with and learn from each other, if science is to be truly rigorous and objective.

But let me, briefly, for the benefit of the others, recapitulate the course of our whole argument, Scientist, so that the others can appreciate too why the considerations that we have developed lead to the view that only person-centred science – or Pygmy science – can be truly rigorous and rational.

We may have rather gone on and on at times, Scientist, but in essence our whole argument was extremely simple.

We began (you will remember) with *standard empiricism*, a philosophy of science which declares: The basic intellectual aim of science is simply to improve knowledge of factual truth. Once we accept this as accurately representing the proper intellectual aims of science, and once we accept that we cannot achieve knowledge about the world independently of experience, then we are led remorselessly to the view that, in order to realise the aims of science, theories and results must be accepted solely on the basis of impartial, unbiased experimental testing. The intellectual domain of science – we are forced to conclude – must be completely dissociated from the domain of feelings, desires, values, personal and social needs, problems and aspirations. It is essential to maintain this decisive split between the *intellectual* and the *personal* and *social*, if science is to deliver authentic, objective, factual knowledge – thus being, in the long run, of benefit to humanity.

Confronted by this immensely influential philosophy of science of *standard empiricism* we ask a simple question: Why? Why do we try to realise the aims for science specified by *standard empiricism*? Why do we seek to improve our knowledge of factual truth?

The answer? We seek factual truth because what we *really* want to discover is *valuable factual truth*, truth that we judge to be in some way interesting, significant, meaningful, important, beautiful, useful.

At once, it becomes clear that the assumptions we make concerning what aims science ought to pursue must be both immensely influential and profoundly problematic. If science is to be truly rigorous, objective and rational, it is essential that this whole topic of aims be thrown open for discussion and exploration, both by scientists and by non-scientists. We are led to adopt the new ideal for science, the new methodology, of *aim-oriented empiricism*, as constituting a more rigorous conception of science than *standard empiricism*.

Again, we ask the simple question: Why? Why do we seek to discover humanly valuable truth?

The obvious answer is that we seek to improve knowledge of valuable truth in order to make such knowledge available to people, so that they may make use of it in order to enrich their lives. We seek valuable truth as a means to the end of helping to promote human progress, enhance the quality of human life. Once again, as a result of this obvious answer, our conception of science is profoundly changed. Knowledge acquires a vital personal, social aspect. Intellectual problems become an aspect of personal, social problems, arising as a part of our concern to solve personal, social problems. Once again, we are led to uphold an even more rigorous ideal for science, the ideal of *humane aim-oriented empiricism* (or *science for people*).

The third step arises from asking the question: can the *methods* of science be profitably applied to life? If we take *standard empiricism* as our starting point, nothing very helpful results. If, however, we take the more rigorous ideal of *aim-oriented empiricism* as our starting point, the situation is profoundly changed: as we saw, this more rigorous methodology can be generalized to form a new, more rigorous ideal for reason – *aim-oriented rationalism* – of universal use and value for all our diverse human pursuits. *Aim-oriented rationalism* offers the hope of enabling us to get into *life*, when judged in human terms, something of the progressively successful quality that is so strikingly found within science, on the intellectual level. *Aim-oriented rationalism* makes clear that life, and philosophies of life, need to be put at the centre of rational intellectual attention – the natural and biological sciences simply developing the more objective, factual aspects of our philosophies of life. Aim-oriented rationalism further provides us with a methodology for developing and appraising our philosophies of life – and so our lives – in directions in which we really want to develop them.

But once we accept all this, we cannot conceivably retain the idea (embodied in *standard empiricism*) that science is essentially an enterprise pursued by experts. Science, intellectual enquiry, has become too intimately associated with our lives, with the very fabric of our beings, for it to be possible to leave it in the hands of experts. We cannot allow experts to determine for us what we shall be, and how we shall live. Once we accept the central idea that emerges from *humane aim-oriented empiricism* and *aim-oriented rationalism* - namely that the fundamental aim of science is to improve our relationships between ourselves, and between ourselves and Nature – then it becomes quite clear that science cannot conceivably realise this basic aim in a truly intelligent, sensitive, effective fashion (i.e. in a rational, rigorous fashion) unless *we* participate in, and help to create, science. We cannot allow the experts to arrange for us what our relationships should be with the world, and with each other. If we take a purely passive role, then *of course* our relationships will suffer, both with the world, and with each other. Expert scientific knowledge is important: but the most important thing of all is that which exists within our own minds and hearts. It exists in our *relationships* with the world, and with each other, and a truly rigorous, rational science would arrange itself, delicately and sensitively, around such relationships in order to help promote their flowering, their enrichment. And for this to work, it is essential that *we* take an active part. Instead of dissociating the intellectual domain of science from our personal relationships, we need to develop a tradition of *using* the intellectual domain of science in order to promote the flourishing of our relationships.

Thus, it is important that we do not allow ourselves to be completely intimidated by the vast stores of esoteric scientific knowledge that the experts have accumulated over the years: and it is important that we are not wholly dismissive of this vast store of knowledge (two attitudes that are often closely associated with one another). Experts and non-experts need to share science. The experts need our help in creating a simpler science, a more human and humane science, a science more in touch with personal feelings and desires – and thus a more *rigorous* science. Left to their own devices, experts are almost bound to lose sight of the human point of it all. The science that they produce can no longer properly realise its human goals. On the other hand, we too have much to learn from experts; expert investigations and discoveries are invaluable in enabling us to extend and deepen our experience, understanding and appreciation of aspects of the world around us, and in enabling us to develop humanly desirable technology. It is only if science becomes a community affair, expert scientific discussion being rationally integrated with general community discussion of community affairs, that science can really help to make the world a happier place.

(The Philosopher pauses; and then suddenly grins delightedly.)

And what I would like to do now is actually to *do* what I have been talking about doing for so long.

239

ROMANTIC: (Not entirely understanding what our Philosopher has been talking about doing for so long.) And what is that?

PHILOSOPHER: I would like to *do* Pygmy science. For a laugh. I would like to sing a song designed to "help make the world happy". A song just for this occasion. For all of us here. It won't take long, I promise. Would that be all right?

(But just at this moment a knock comes on the door.)

PHILOSOPHER: (Irritated) Come in!

(And now something rather peculiar happens. Who should open the door – somewhat timidly – but the author himself. It is, I admit, somewhat unusual for an author to appear in his own book, but so many conventions have already been broken that I don't suppose it will matter if this one is broken as well.)

N.M: (Shyly) Hullo.

(Cries of protest and outrage. Quite naturally, fictional people are not at all used to having to cope with an author – whoever he or she may be – intruding into their lives in this kind of way.)

N.M: I am very sorry to disturb you all like this. It's just that I felt that perhaps *I* should sing this song to help make the world happy – a song for us here and now, on this occasion, at least.

REBELLIOUS ROMANTIC: It's bad enough with *one* philosopher around. But two of them! You are a philosopher, I suppose?

N.M: Well, yes, I suppose I am, in a way.

LIBERAL: But what exactly is this "singing in order to help make the world happy" idea anyway? It sounds to me just a little – how can I put it? Far fetched? Implausible? Pretentious?

N.M: What I had in mind was to draw a kind of thumb-nail sketch of an ideal world, a perfect world, in which there would be perfect harmony between ourselves, and ourselves and Nature. What I would like to do is to take the materials that we have to hand – all your diverse philosophies of life – and stitch them together as best as I can to form a new coherent whole. Not that I expect you to abandon your own philosophies of life, or anything like that. It is merely that I thought it would be fun to see whether a coherent whole can be put together from these diverse elements – for a laugh, as our Philosopher here said. As an example of bubble blowing and bubble bursting.

(There is a kind of confused murmuring of objections, encouragement to go on, and dissent. I decide it will be all right if I do continue.)

Imagine that you cup your hands, and before your eyes, step by step, layer by layer, a universe is created, like an onion. Each layer will be made from what seems to *me* to be the very best in all your diverse philosophies. But just because I have to tuck these various layers together to make a nice smooth, rounded onion, I am afraid that you may find some of your most precious ideas somewhat folded and distorted out of their proper shape. This is, I am afraid, almost inevitable. It's very difficult to tuck in all the layers so that they all fit smoothly together to form a sensible onion. But I promise I will do the best I can.

CHRISTIAN: Is it not a little bit disrespectful to be talking about an *onion* in these circumstances? If I understand you properly, you are about to share with us your vision of Ultimate Reality, Paradise, Utopia, Heaven on Earth. You are about to describe what would, I suppose, be called "The Form of the Good" if this were a Platonic dialogue. Surely the humble *onion* does not quite match up to these grandiose ambitions?

N.M: First, I am delighted with your account of what I am trying to do here. My concern is to draw a rough sketch of a possible universe which is as beautiful, as perfect, as humanly desirable, as I can imagine, and which, at the same time, is as realistic, as adequate a representation of facts, as I can manage. My concern is to draw the most realistic, humanly beautiful world that I can imagine. It is just this, according to aim-oriented empiricism and aim-oriented rationalism that constitutes the heart of scientific enquiry. With as sensitive an attention to reality as we can muster, we seek to divine the hidden beauty of the world – that which can bring joy to our hearts. We seek that which can make the world happy.

Why do I employ the emblem of an *onion* for the most perfectly beautiful universe that I can imagine? I think it is because I do not wish to be taken too seriously. Bubble blowing is a game that anyone can play: it is as easy as slicing up an onion – or rather, as difficult as putting together again a sliced-up onion. And besides, my concern here is only to indicate a possible way in which what seems to me to be the best in all your various philosophies can be fitted together to form a harmonious whole: all the rich, beautiful details will be missed out altogether. All in all, an onion is probably as good an emblem for this strange, rich, beautiful, disturbing world in which we find ourselves as my "vision" – as you call it – will turn out to be.

Since our concern here has been primarily with *science*, let us begin with the contribution from science and, in particular, from theoretical physics. Here I simply take over almost wholesale Einstein's vision: Everything that occurs in the universe does so in accordance with some simple, unified, harmonious *pattern*. Everything is ultimately made up of just a very few different sorts of fundamental physical entities – perhaps in the end of just *one* entity, something rather like the unified field Einstein spent his later years seeking to divine. A precise, mathematically simple, unified, coherent theory specifies the way in which these entities interact, or the way in which the one entity changes its state in space and time. One might call this doctrine *Pythagorianism*. The only slight disagreement that I would have with Einstein would be that I would not wish to commit Pythagorianism to *determinism*: it may be that the fundamental entities interact with one another probabilistically, rather than deterministically – as quantum theory might appear to suggest.

SCIENTIST: It is absurd to suggest that this vague, rather waffly, metaphysical stuff has anything to do with physics, or with science. In order to qualify as a *scientific* idea a theory must at least be experimentally testable –

PHILOSOPHER: (Breaking in) But that ignores completely our earlier discussion about *aims*. Surely we agreed that our metaphysical conjectures about the vast domain of all that of which we are ignorant, but hope and desire to discover, ought to be regarded as being just as vital a part of our scientific knowledge as our best testable theories? Something like Pythagorianism is clearly in practice presupposed to be true by science, in that theories which clash too violently with this doctrine are simply never *formulated*, let alone considered, or put to the test. Do not honesty, rigour and objectivity demand that such influential, implicit presuppositions be made explicit?

SCIENTIST: It is a grotesque misrepresentation of science to suggest that Pythagorianism, as you call it, is actually a part of scientific knowledge.

PHILOSOPHER: But –

N.M: Perhaps we could just agree that it is from the standpoint of physics a beautiful metaphysical conjecture passionately upheld by many theoretical physicists, it being possible to interpret much of the progress of theoretical physics as a steady movement towards the articulation of Pythagorianism in a testable form. Pythagorianism is, as it were, the ideal state of affairs, as far as theoretical physics is concerned.

SCIENTIST: This doctrine of Pythagorianism is so vague, so open to diverse interpretations, that I suspect that, in the end, it probably asserts nothing definite about the world whatsoever.

PHILOSOPHER: On the contrary: it denies the existence of such things as ghosts, Cartesian minds interacting with brains, poltergeists, psychic phenomena, life after death, magic, witchcraft. This, surely, is evidence enough that something substantial is being asserted.

SCIENTIST: But what of biological phenomena?

N.M: That comes next. Biological phenomena constitute the next layer of the onion. In a sense, of course, biological phenomena are nothing new; they are just physical phenomena, and can in principle be understood in terms of the basic physical laws, governing all physical phenomena. The distinctive feature of biological systems, however, might be put like this: groups of fundamental physical entities have come to be arranged in highly distinctive patterns, having the capacity to perpetuate such distinctive patterns, as long as certain physical conditions remain constant (the sun does not blow up, etc.). Life, one might say, represents a new kind of pattern superimposed on the basic physical pattern of Pythagorianism. This new kind of pattern depends upon the perpetuation of a pattern in *initial conditions*, as physicists would say. To put it crudely: we are alive because of the maintenance of an incredibly rich and complex pattern of physical entities that go to make up our brains: destroy that pattern and we are no longer alive.

Biological phenomena can be treated as an aspect of *physics*. In this case, biological phenomena are just very complicated, unusual physical phenomena, amounting to no more than the maintenance and evolution of highly unusual *initial conditions*, everything of course occurring in accordance with the basic physical laws.

We may, on the other hand, decide to do *biology* rather than physics: that is, we may choose to study biological phenomena as *life*. In this case, we study aim pursuing entities; we study ways of life, and the physical mechanisms underlying life. In particular, our concern with the evolution of life leads us to study the way in which changes in ways of life lead to changes in physiology –

SCIENTIST: But that is absolute nonsense. It amounts to nothing but the most preposterous, outdated Lamarckianism. If modern biology teaches us anything, it teaches us that acquired characteristics are *not* inherited.

N.M: Agreed. I do not at all wish to imply that acquired characteristics are inherited. The point I wish to make is rather different – but one, nevertheless, which modern biologists have been reluctant to acknowledge, just because it smacks of discredited Lamarckianism.

It can be put like this. Imagine a dog-like species, living near rivers, but hunting fast-running land animals. Let us suppose there is a mutation (a chance physical occurrence) producing a puppy with flippers instead of legs. Clearly in these circumstances, the chances of the puppy surviving are almost non-existent; the puppy will not live to breed, and perpetuate beaver-like dogs. Suppose on the other hand that the animals which the dogs hunt die out (due, perhaps, to climatic changes), and the dogs discover that fish can be caught and eaten – the discovery initially, doubtless, being largely accidental, as many discoveries are. Relative to this changed way of life, flipper-type mutations will have immense survival value. A mutant puppy, with flipper-type legs, will now have a very good chance of surviving and breeding. In order to understand the development of the beaver-type animal that emerges as a result of these successful mutations, at least two causes will need to be attended to: (1) the change in the way of life, and (2) the mutation. The development of flippers is, we may say, in part a long term, unintended, consequence of the "decision" of the dog-like animals to catch and eat fish. And more generally we may accurately say that we are ourselves here in part as a kind of long term unintended consequence of millions upon millions of animals pursuing their lives with tenacity, intelligence, liveliness, courage. We would not now exist if countless generations of animals had not displayed these qualities in their lives. We have a great deal to thank animals for.

If you don't believe any of this, then please go off and read Alistair Hardy's beautiful book, *The Living Stream.*[1] In a moving, simple and, in my view, wholly convincing way, he sets out this profoundly important new version of the theory of evolution – this almost Lamarckian version of neo-Darwinism – with a wealth of fascinating examples.

SCIENTIST: So you believe everything occurs in accordance with physical law, and yet you are an anti-reductionist?

N.M: Exactly. In considering evolution, incidentally, one can perhaps distinguish some four or five stages in which ways of life are perpetuated and developed. (1) Behaviour is almost entirely determined genetically, so that behaviour is the outcome of genetics and environment. (2) Behaviour is to some extent taught, learned, acquired by imitation, so that important *biological* changes can occur – namely, changes in ways of life – having important subsequent morphological consequences, even though there are no genetic changes (as illustrated by our mythical dog into beaver story). In other words, learning, education, culture, ways of life, have become essential *biological* concepts. (3) There

245

is the development of language, which enormously enhances the capacity of the animal in question to act intelligently and cooperatively. An animal without language is obliged physically to try out actions in attempts to solve problems (some of which may end in death). With the development of language and imagination, however, actions can be tried out in imagination, their success or failure being imagined, so that all kinds of possibilities can be explored – even cooperatively – without any actual action being performed at all. (4) There is the development of enduring cultural artefacts – writing, books, etc. – which greatly increase the capacity of the species to hand down possible ways of life – ideas and discoveries relevant to various ways of life – from generation to generation. (5) Finally, there is the possibility of a species consciously arranging for the pursuit, the encouragement and development, of that which is of greatest value in life, by applying to life aim-oriented rationalist ways of thinking, so that, for example, consequences of actions are, as far as possible, foreseen and intended, rather than being heedlessly ignored. A species takes responsibility for itself, and intelligently and effectively arranges for the flourishing of that which is of greatest value in life. Needless to say, this has not yet come into existence.

REBELLIOUS ROMANTIC: (Absolutely furious) But what on earth makes you think you are entitled to speak of "that which is of greatest value"? The view of things that you have presented to us denies utterly that there can exist anything of value. Everything is physics. There is nothing more than physical entities twirling around in accordance with physical law. Reality is just a gigantic machine. There can be no *life* in your universe, no *passion*, no *imagination*, no –

N.M: Your objection brings us straight to the *third* layer of our onion – the domain of humanly significant qualities and attributes. The world is charged with significance and beauty. The colours we see, the sounds we hear, the smells we smell, really are out there in the world. The sky really is blue – intensely, dramatically blue, at times, straight above, on a Spring day, or delicately blue-green, near the horizon, at evening, over the sea. In the world there are things, scenes and events that are beautiful, tragic, ugly, horrible, moving, courageous, delightful, dull, painful, glorious, endearing, lovable; there are people, and actions, imbued with intelligence, stupidity, sensitivity, brutality, kindness, warmth, honesty, grace, delight, suffering, madness, sanity, on and on. You don't need to be told these things. All kinds of extraordinary beautiful, wonderful, moving things *exist* in the world. Out there, objectively, independently of ourselves. Painters such as van Gogh, let us say, or Renoir, explore aspects of reality just as *real* as those aspects explored by scientists: and so too poets, novelists, dramatists, sculptors. And above all, *we* explore aspects of reality in our lives, in terms of our own personal experiences, emotional reactions, ideas and values. Not everything we think we see actually exists by any means. Just as we can see visual hallucinations, so too we can experience emotional and evaluative hallucinations; but that does not mean that *all* our personal experiences are systematically delusive, the outcome of illegitimate projections from "subjective experience" into the objective domain – like a red after-image projected onto a white surface.

REBELLIOUS ROMANTIC: I believe all that. But *you* can't believe it. You said that *everything* is made up of fundamental physical entities. For you, *everything* is physics. The mad, single vision of science.

N.M: Theoretical physics, yes, does seek to develop a comprehensive view of reality. But only a comprehensive view of a highly specialized, highly restricted type. The task of theoretical physics, we may say, is to discover the *universal* pattern in events – that aspect of things which everything has in common with everything else. Theoretical physics seeks to select out the very smallest possible number of properties of things which everything has in common with everything else – fundamental physical properties analogous to momentum, angular momentum, electric charge, and so on, of current physical theory. The task of physical theory might be put like this: To select out a few basic properties, P_1, P_2,.....P_6 say, which are such that any possible isolated system whatsoever can have its state described in terms of these properties – this description of initial conditions, together with the fundamental physical theory, sufficing in principle to imply true

247

descriptions of subsequent states of the system, in terms of the values of the basic properties $P_1,.....P_6$. Theoretical physics, we may say, is concerned only with universal, causally efficacious properties. All other kinds of properties possessed by things are completely ignored. A completed physics would, in principle, apply to all that there is, but it would be no means tell us all that there is to know about all that there is. Theoretical physics explains, renders intelligible, various natural phenomena by revealing a common, underlying pattern in ostensibly diverse phenomena: and the price that has to be paid for rendering phenomena comprehensible in this kind of way is that one ignores what is individual and distinctive about things. For example, theoretical physics would describe people as if they were not essentially different from stars, crystals or bricks: there is merely a different number of fundamental physical entities arranged in a different way. What is distinctive and extraordinary about people, all that which makes us *people*, as opposed to merely physical systems, in no way essentially different from stars, is just ignored by theoretical physics, as being irrelevant to the *aims*, the concerns, of theoretical physics. Physics may explain why light of such and such wavelength is reflected and absorbed by grass; but physics will never explain why grass is *green*, as experienced by us, just because that is the kind of property theoretical physics is obliged to ignore in seeking to discover a simple, universal pattern in all phenomena. Greenness as experienced by us will never be detected by a scientific instrument: in order to detect greenness you need to be a *person*. Physicalism provides us only with a kind of skeleton view of the universe: all the flesh, the colour, the drama, the beauty, the passion, the human significance, is left wholly out of account. Theoretical physics does not *deny* the existence of the flesh, the colour, the drama, the passion, the human meaning: nor does theoretical physics in any way imply that this aspect of the world must, as it were, be scraped off material objects, physical processes, and be tucked into the domain of Mind, Consciousness, associated somehow with human brains, in the kind of way presupposed by Descartes (and so many other thinkers who have taken Cartesian dualism for granted). The simple fact is that theoretical physics seeks only to provide a complete description of a highly selective *aspect* of reality. All that exists which is not mentioned within theoretical physics is ignored simply because theoretical physics can realise its goals without needing to mention all these additional features of things – the goals of theoretical physics indeed only being realizable if these additional aspects of reality *are* ignored.

SCIENTIST: So the sky is blue even if there is no one around to experience its blueness – and it would be blue even if, as a matter of fact, no one ever existed to observe its blueness?

N.M: Yes. Or to take your point even further, there are, in the world all around us, all kinds of properties and features of things which we can never be aware of, just because we do not possess the appropriate kind of nervous system and sense organs. The world is far, far richer in qualities and properties than we can ever be aware of. The world as we ordinarily experience it amounts, in some ways, to an even more highly selected aspect of what there is than the viewpoint of physics.

One of the ways in which, in my view, all this constitutes a distinct improvement over many traditional Romantic ideas, influencing Romantic thought and art, is that it avoids the extreme subjectivism that tends to be associated with Romanticism. We need to take our personal, emotional/sensual experiences seriously, and thoughtfully, as possible indications of the nature of aspects of reality, aspects of what really is out there, in the world. We need to develop emotional reactions accurately responsive to objective reality. Traditional Romanticism tends to lead to the absurd, and often harmful, conclusion that emotions are simply to be *expressed*, vented upon the world – just because the possibility that emotions might be fallible pointers to the nature of aspects of objective reality is ignored.[2]

(The Rebellious Romantic is about to protest when the Christian intervenes.)

CHRISTIAN: But how can life have any meaning if there is no God? As far as I have been able to understand it, there is no place in your scheme of things for spiritual realities – for faith, worship, God, the soul, salvation.

N.M: Your question brings us straight to the *fourth* layer of our onion. Christianity has a profound contribution to make in emphasizing the need for faith, for trust, for love. Science cannot, for example, ever prove that a beautiful mathematical fugue runs through all natural phenomena: in the end such an idea must rest on faith, on trust. Even if a unified physical theory is one day formulated, which meets with immense empirical success, nevertheless –

CHRISTIAN: But you don't understand. What I am protesting about is the lack of anything in your ideal world which can inspire faith and trust as far as ordinary people are concerned. What you have described may be a possible ideal for theoretical physicists, for biologists and for artists: I do not know. But as far as I can see you have described nothing which can constitute an adequate substitute for faith in a personal God who cares for, and loves, humanity, and each one of us individually.

N.M: Well, I must just confess, I suppose, that the kind of universe that I have described seems to me more desirable, in human terms, than the world of Christian belief.

CHRISTIAN: How so?

N.M: For me, the fundamental religious problem that arises within the Christian faith is not: How can God forgive us? But rather: How can we forgive God? God is, I gather, omniscient, omnipotent and omnipresent. This must mean that God acts on us through natural phenomena. But natural phenomena maim, torture and kill people, day after day, every day, every hour. In the end, we are all killed by natural phenomena. Even when a person dies by the hand of another, natural phenomena invariably intervene. All this must mean, then, that God maims, tortures and kills people. I find it very hard to understand how a God that does this can be a God of love.

CHRISTIAN: You must realise, of course, that the problem you indicate has received a great deal of attention within Christian theology.

N.M: Yes, I do realise that, although I must admit to being largely ignorant of the orthodox discussion of this problem. It seems to me, however, that it is very easy to see that there can only be, two possible solutions to the problem. Either God does not know what He is doing, and cannot know; or God knows and understands what is going on, and is doing all that He can to prevent what is happening, but unfortunately lacks the capacity to be very effective in the world. I believe in both solutions.

The "God" of the first solution is Einstein's "Old One". Poor Nature; She tortures us and kills us, but only because she cannot know and understand what it is that She does to us. If She were capable of understanding, She would at once cry, "Oh, people, forgive me, forgive me, I did not know what I did to you". But just because She cannot know and cannot understand, She needs our active help to ease Her heedlessly cruel treatment of us; She needs to be helped to treat us more kindly. There is, in other words, I believe, a profoundly heartening religious message locked up in theoretical physics. Nature can be forgiven – Nature can even be loved – for She cannot know what She does to us, and hence cannot be blamed.

The "God" of the second solution is something like "that which is best, potentially or actually in each one of us", "that which feels with others, feels for the suffering, the cares, of others, but which is not always capable of acting – for one reason or another".

For me, science is a profoundly religious enterprise. The aim of the natural sciences is to seek, and to disclose to us "God" in the first sense. The aim of the social sciences is to help us to discover, develop, strengthen, "God" in the second sense. Is this a universe of love? The issue is up to us. *We* need to take responsibility for, and care for, our world.

CHRISTIAN: But how can people be expected to act selflessly if there is no God external to us who cares for us, and loves us?

N.M: As to that, I am not at all sure that people ought to act selflessly. If I were to sum up my philosophy in three sentences I think I would do it in terms of the following Talmudic saying:

If I am not for myself, who will be for me?
If I am for myself only, what am I?
If not now – when?[3]

What I particularly like about this saying is that it comes in the form of three questions. Nothing is asserted; we are left to answer the questions ourselves, as best we can.

I find myself giving different answers to these questions as time goes by; at the moment, my answers would go like this.

If I am not for myself, then many people will be for me – my family, my friends, my acquaintances, even strangers that I meet. But no one can be as perceptively, as sensitively, as lovingly for myself as I can be – just because no one is so intimately associated with myself as I am. Of all humanity, it is myself that I should be for first.

If I am for myself only, what am I? I am not sure. I seem to be such different things on different occasions. At times it seems to me that when I am most exclusively for myself, I experience richness and beauty. "No sooner am I alone than shapes of epic greatness are stationed around me."[4] At other times, it seems to me that when I am most exclusively for myself I become delightfully, wonderfully small, fragile, ineffective, ordinary, bumbling. It can be a delight to take up so little room, to be so sensitively fragile as to be scarcely noticeable – to be just that which at times I have sought to flee from as being altogether too insignificant and ineffective to be *me*. Certainly it seems to be desirable for me to be on occasions for myself alone, allowing everyone else, the whole of humanity throughout all time, the whole cosmos, to lapse into insignificance.

But how enriched I am by the existence of others; and my being is involved with the being of others: in being for them I am being for myself as well; and in being for myself I am being for them. If I do not manage to be for others very successfully, well, the same goes for how I am for myself.

If not now – when? Well, of course, all this goes on for all of us all the time – since the day we were born. It is only a question of how successfully one puts it into practice.

CHRISTIAN: I find your answers almost horrifying. You seem to be recommending that we should be – how shall I put it? – qualified psychopaths.

N.M: I am sorry that you are horrified. I suppose I believe that one of our troubles today is that many of our values and ideals, inherited from the past, make us feel *guilty* about being "qualified psychopaths" as you put it. It seems to me that in an ideal world we should all, in a quite easy and natural way, be qualified psychopaths in a non-guilt-ridden fashion. It is just when one is wholeheartedly for oneself that one is released to be for others as well, as an extension of being for oneself. That is how I see it, at any rate. Calls for self-sacrifice always seem to me to be a great mistake.

MARXIST: All this is all very well; but if we consider for a moment the brutal realities of the world today, all your talk becomes so much hot air. A moment ago you were talking of taking on responsibility for our world. But that is exactly what we *cannot* do, as long as we are all enmeshed in the capitalist system. Only by working for a complete transformation of our whole economic order can we hope to make any kind of contribution towards creating a world in which all of us can take up our share of responsibility for our world.

N.M: Your comment brings us to the *fifth* layer of the onion. I know almost nothing of the writings of Marx: so please forgive the ignorance of my remarks. But it seems to me that there are at least three points of great value to be extracted from Marx. First, there is his central concern for the problems of the oppressed, the poor, the exploited. Second, there is his concern to apply science to social and political problems – problems experienced by people in their lives – in order to help solve these problems. And third, there is his emphasis on the *economic* aspect of our activities and endeavours – an aspect we are often all too ready to forget, partly, perhaps, as a result of the Christian habit of dividing reality up into the "spiritual" on the one hand, and the "material" on the other hand.

Now for my disagreements. In the end, I believe, we must learn to live with one another, as harmoniously as we can, encouraging, or at least tolerating, diversity. Marx, however, seems to advocate, or at least prophesy, the victory of the working class over other classes, leading to the annihilation of the other classes. To my ears, this amounts to advocating tribal war – even tribal genocide.

Second, Marx's attempt to apply science to help solve our social problems suffers in my view from two defects. He upheld a seriously inadequate conception of the natural sciences; and much more disastrously, he made the mistake that is made by almost all other social scientists, right down to the present day, of assuming that the proper task of the social sciences is to improve our *knowledge* of social phenomena, on analogy with the natural sciences improving our knowledge of natural phenomena. As our Philosopher here has just argued, a far more rational and humanly desirable task for the social sciences is to work out how to apply the *methods* of the natural sciences to our various human, social pursuits. The most elementary requirement for rational problem-solving is to articulate the problems to be solved, consider alternative possible formulations, develop and assess possible solutions to the problem. *This* is what we need to do if we are to tackle our human social problems in a rational, intelligent and humanly desirable fashion. It is *this* which is advocated by aim-oriented rationalism. And yet it is precisely these obvious points that are lost sight of by orthodox approaches to the social sciences. (The fundamental task of sociology, for example, becomes to solve *sociological* problems, problems of knowledge and understanding, new knowledge subsequently being applied, perhaps, to help solve *social* problems. All is reversed by aim-oriented rationalism: the rational, rigorous place to begin is with the articulation of *social* problems – problems experienced by people in their lives.) Marx too makes precisely the same mistake. For him, the search for knowledge comes first; then comes the question of what action one can hope to perform within the confines of the socio-economic laws that one has discovered governing the development of society. This approach, of itself, is bound to lead to the conclusion that we have only limited freedom, limited room to manoeuvre. The more successful is our search for laws governing social development, the more limited, we will be obliged to conclude, is our freedom. The whole Marxist (and orthodox) approach to the social sciences has built into it an aim which must work against our best interests.

And of course the kind of approach to social problems advocated here would be condemned by Marx out of hand as "Utopian".

Nevertheless, it seems to me that aim-oriented rationalism does answer to the deepest aspiration of all in Marx's work – namely, to apply science (or the methods of science) to the task of helping to solve our human, social problems.

Again, the emphasis that Marx gives to the economic aspect of life is important. But in my view he goes disastrously wrong when he argues that the economic aspect *determines* the rest. The mistake is to *dissociate* the economic aspect, from moral, intellectual, cultural, political and legal aspects of life. According to aim-oriented rationalism, we need to represent our aims accurately whatever we may be doing: and insofar as our aims have an economic aspect, then it is important that this is represented. For example, science cannot be properly understood without some understanding of the economic aspects of science.

It seems to me to be a hopeless oversimplification to lay the blame for all our troubles on the capitalist system. There even seems to me to be the germ of a highly desirable idea within the free enterprise system. In an ideal world, it ought to be possible for individuals to produce goods, on their own initiative as it were, that are of use or of value to others – resources being available for people with good ideas, and the capacity to put them into practice. Individual initiative is important, if we are to avoid all becoming institutionalized. The trouble is that through a lack of what might almost be called *moral* attention, the free enterprise system has been allowed to develop in humanly undesirable directions. I believe, quite generally, that, by and large, we need to try to develop what already exists in more humanly desirable directions. Why should we not try to create a loving free enterprise system?

MARXIST: A loving capitalism! I have never heard of anything quite so obscene. How can you come out with such things?

N.M: I am sorry.

MARXIST: I find your critique of Marxism wholly unacceptable, and entirely uninformed concerning Marx's actual views. But let that pass. If the capitalist system is not to blame for our troubles, what, then, in your view, is responsible?

255

N.M: To begin with, I am, I suppose, wholly unsurprised by the existence of problems. You see, I really do believe that, whatever else we may be, we are at least natural phenomena, subject in all our actions to the laws that govern natural phenomena. The miracle is that we have done so well, and achieved so much. I see our capacity to achieve humanly desirable ends as extremely precarious. Four circumstances in particular have, I am inclined to suppose, added to our difficulties. (1) Unlike other animals, we alone, I imagine, are aware of our inevitable death. Much of our science, culture, religion, myth-making, may in fact be an attempt to solve the problem of death. Our awareness of our inevitable death plays havoc, I am sure, with our intuitive animal/emotional responses. (2) Our emotional make-up was, I suspect, developed to suit so-called primitive, tribal life. Subsequently, we have transformed utterly the conditions and circumstances of our lives, without always informing our environment of our feelings, and our feelings of our environment. A gulf has developed between the objective and the personal, the factual and the emotional, as our Philosopher here has argued. (3) It is possible that, built into our biological/emotional make-up, there is the need to identify with a relatively small tribe of close associates, in opposition to other, "hostile" tribes. This may simply be an evolutionary device designed to disperse mankind across available land. In our modern world, however, this reflex – if that is what it is – creates the most appalling problems, and all but threatens to destroy us all. (4) Not surprisingly, our ideas, our values and ideals, our *aims*, have not always been of the best. Again, as our Philosopher here has argued, this is inevitable: it is inherent in the very situation of pursuing aims in the real world. Hence the importance of aim-oriented rationalism.

RATIONALIST: It is at this point, I imagine, that my viewpoint is supposed to be incorporated into your onion?

N.M: Yes.

RATIONALIST: I find your whole conception of Reason entirely unacceptable.

LIBERAL: What about my viewpoint?

N.M: Well, could it not be said that what I have been advocating is a theology of *persons*? Liberalism seems to me a basic component of the viewpoint that I have been advocating.

PHILOSOPHER: What about the Buddhist's viewpoint?

N.M: I suppose I would say that one thing that we may seek to do in life is to rediscover the *animal* within ourselves. I am enormously impressed by the vitality, the liveliness, the courage, the tenacity for life, exhibited by animals. Think of a mouse living in a modern city. No National Health, no social workers, no social security, no psychiatrists, no shops, banks, experts, police, no science, no art, no television, few comforts, and an appallingly hostile environment. In such circumstances, most of us would be quite unable to cope. And yet the small, defenceless mouse seems to cope very well – even though there has been no education, no training. We are animals too! And we can, I think, discover within ourselves something of the same tremendous optimism, courage, liveliness. We have, I think, a fierce impulse for life, a passionate desire to take part in, to share, something of value, of beauty, during the course of our lives – we do not know quite what it is. The knowledge of our impending death, this being our only chance, makes the whole thing more urgent. I think we do not take this simple, basic animal impulse very seriously in our modern world. On the one hand, it tends to be associated, in many people's minds, with discredited religious viewpoints. On the other hand, various unsatisfactory *substitutes* for this kind of impulse are pursued in our modern world. We do not apply aim-oriented rationalism to our basic impulse to live!

BUDDHIST: I am afraid I can find almost nothing of my views in what you have just said.

N.M: Scientist? Surely *you* have a certain sympathy with my onion?

SCIENTIST: I am afraid I must disappoint you. Pythagorianism seems to me to be a metaphysical viewpoint only rather loosely associated with science. Your interpretation of biology, and the theory of evolution, I find entirely unacceptable.

257

N.M: Liberal, surely you have some sympathy with my "song to make the world happy"?

LIBERAL: It is a little too fanciful for my taste, I am afraid. I prefer to put my trust in common sense.

N.M: Philosopher? You, at least, I am sure, will not desert me in my hour of need.

PHILOSOPHER: I am afraid I find your presentation of your yews extremely unsatisfactory. I do not think you have succeeded in solving the mind-brain problem; you have said nothing concerning the problem of free will. In several of my as yet unpublished papers I deal with these, and related, problems in a much more satisfactory manner than anything that you have achieved here.

N.M: Oh clear. My song to help make the world happy, to help bring a little more harmony amongst ourselves, and between ourselves and Nature, seems to have achieved only disagreement and discord.

WINO: Here, I wouldn't call that much of a song.

[We are all amazed. We have forgotten the Wino. He now gets to his feet, clutching a bottle in one hand, and he begins to drone some appalling song that none of us can recognize. The Rebellious Romantic recognizes a kindred spirit: he joins in. And in no time we are all singing, making the most extraordinary noise, and roaring with laughter at our ineptitude.]

THE END

NOTES

Chapter 1

1. Colin Turnbull, *The Forest People*, Picador, 1976. I first learned of the Pygmies of the rain forests of central Africa from Brian Easlea, who has, on a number of occasions, movingly portrayed the open, loving, trusting way of life of the Pygmies, a way of life that is in such striking contrast to so much of our modern "civilized" world.
2. See Henry Troyat, *Tolstoy*, Penguin Books Ltd., 1970, p.73.

Chapter 3

1. W. I. B. Beveridge, *The Art of Scientific Investigation*, Heinemann, 1961.
2. Albert Einstein, "Reply to Criticisms" in P.A. Schlipp (ed.) *Albert Einstein: Philosopher Scientist*, Open Court, 1970, pp. 683-4.

Chapter 4

1. Even as he recites these lines by Yeats, an awful pang of doubt shoots through our Philosopher's heart. "Oh dear," he thinks, "I seem to spend my time hovering in an agonized way somewhere between lacking all conviction and being full of passionate intensity. Does that make me one of the best, or one of the worst?"
2. G. Ryle, *The Concept of Mind*, Hutchinson, 1949.

Chapter 5

1. For a formulation of a somewhat similar viewpoint see N. Maxwell, "A New Look at the Quantum Mechanical Problem of Measurement", *American Journal of Physics*, 40, 1972, pp. 1431-5; and "Towards a Micro Realistic Version of Quantum Mechanics", Part 1 and Part 2, *Foundations of Physics*, 6, 1976, pp. 275-92 and 661-76.
2. Quoted in Banesh Hoffman, *Albert Einstein: Creator and Rebel*, Hart-Davis, 1973.
3. W.W. Sawyer, *Prelude to Mathematics*, Penguin Books, 1969, p. 19.

Chapter 6

1. K. Przibram (ed.), *Letters on Wave Mechanics*, Vision Press Ltd., 1967, p. 39.
2. *Ibid.*, p. 31.
3. "Editorial Comment", *Reviews of Modern Physics*, 42, 1970, p. 357.
4. Curiously enough, some of the author's own work and experiences are at this point remarkably similar to the Philosopher's.
5. See E. Wigner, *Symmetries and Reflections: Scientific Essays*, M.I.T. Press, 1970. Also the present author's "The Rationality of Scientific Discovery", *Philosophy of Science*, 41, Part 1, pp. 123-53; Part 2, pp. 247-95.
6. A large flat platform, rotating at immense speed, will exemplify the geometry of the Euclidean plane if measurements are made by someone who is at rest. But if measurements are made by someone who is carried around with the platform, the surface will appear to be non-Euclidean, just because measuring rods will "contract" as they are carried from the centre towards the rim of the platform, as a result of the acceleration that they acquire due to the greater rotational speed of the outer rim. Similarly, clocks will slow down as they are carried towards the outer rim of the platform.

Chapter 7

1. A. Einstein, *Ideas and Opinions*, Souvenir Press, 1973.
2. P.A. Schlipp (ed.) *Albert Einstein: Philosopher Scientist*, Open Court, 1970.
3. M Born, *The Born-Einstein Letters*, Macmillan, 1971.
4. D.H. Lawrence, *Selected Letters*, Penguin Books Ltd., 1971, p. 96
5. A. Einstein, *Ideas and Opinions*, *op. cit.*, pp. 38-40.
6. For a presentation of a somewhat similar line of thought see N. Maxwell, "A Critique of Popper's Views on Scientific Method", *Philosophy of Science*, 39, 1972, pp. 131-52; and "The Rationality of Scientific Discovery", *Philosophy of Science*, 41, 1974, pp. 123-53 and 247-95.

7. Our Philosopher would probably add, if he stopped to think about it, that even more recent writers, such as Kuhn and Lakatos, who have in a sense given a place to metaphysical ideas in science, in the form of "paradigms" or "hard cores", still miss the essential point; they still make the basic standard empiricist assumption that the fundamental aim of science is to discover truth *as such*, obtain empirical growth *as such*. For a slightly more detailed discussion of this point, see the author's "The Rationality of Scientific Discovery", *Philosophy of Science*, 41, 1974, pp. 123-53.

8. Arthur Koestler, *The Sleepwalkers*, Penguin Books Ltd., 1964.

9. The simple point that Hume's essential claim is absolutely correct – this in no way negating the authenticity of scientific knowledge – was perhaps only clearly realised amongst philosophers as a result of the work of Popper.

10. In her delightful biography of the great mathematician, David Hilbert, Constance Reid has the following to say about the "stupidity" of this brilliant man. She is writing about the high-powered graduate seminars held by Hilbert in Göttingen during the 1920's, and she writes: "The bright newcomers who saw the famous Hilbert in action for the first time at these events were struck by his slowness in comprehending ideas which they themselves "got" immediately. Often he did not understand the speaker's meaning. The speaker would try to explain. Others would join in. Finally it would seem that everyone present was involved in trying to help Hilbert to understand.

" 'That I have been able to accomplish anything in mathematics', Hilbert once said to Harold Bohr, 'is really due to the fact that I have always found it so difficult. When I read, or when I am told about something, it nearly always seems so difficult, and practically impossible to understand, and then 1 cannot help wondering if it might not be simpler. And,' he added, with his still child-like smile, 'on several occasions it has turned out that it really was more simple!'" (Constance Reid, *Hilbert*, George Allen and Unwin, 1970, p. 168.)

11. This is really very embarrassing. I have no idea how our puppet Philosopher got to hear of this story. The "friend" is of course myself.

12. BSSRS no longer exists, but another organization, Scientists for Global Responsibility, with somewhat similar aspirations, does exist.

13. Oliver Sacks, *Awakenings*, Penguin Books Ltd., 1976.

14. Samuel Butler, *Erewhon*, Penguin Books Ltd., 1970.

15. Karl R. Popper, "Reason or Revolution?", *The Positivist Dispute in German Sociology*, Heinemann Educational Books, London, 1976, p. 296.

Chapter 10

1. A. Hardy, *The Living Stream: A Restatement of Evolution Theory and Its Relation to the Spirit of Man*, Collins, 1965.

2. For a somewhat more detailed "philosophical" discussion of aspects of these three layers of the onion, see N. Maxwell, "Physics and Common Sense", *British Journal for the Philosophy of Science*, 16, 1966, pp. 295-311; "Understanding Sensations", *Australasian Journal of Philosophy*, 46, 1968, pp. 127-45; "Can there be Necessary Connections between Successive Events?", *British Journal for the Philosophy of Science*, 19, 1968, pp. 1-25, reprinted in R. Swinburne (ed.) *The Justification of Induction*, Oxford University Press, 1974, pp. 149-74.

3. Quoted in E. Fromm, *The Fear of Freedom*, Routledge and Kegan Paul, 1960.

4. John Keats, in *Keat's Letters: A New Selection*, edited by Robert Gittings, Oxford Paperbacks, 1970, p. 170.

FURTHER READING

A Personal Selection of Books Related in One Way or Another to the Main Themes of This Book Pre 1976

Virginia Axline, *Dibs: In Search of Self*, Penguin Books Ltd, 1971.

Peter G. Bergmann, *The Riddle of Gravitation*, Charles Scribners' Sons, 1968.

W.I.B. Beveridge, *The Art of Scientific Investigation*, Heinemann 1961.

Samuel Butler, *Erewhon*, Penguin Books Ltd. 1970.

R.G. Collingwood, *An Autobiography*, Oxford University Press, 1970.

Tobias Dantzig, *Number: The Language of Science*, George Allen and Unwin Ltd., 968.

Pierre Duhem, *The Aim and Structure of Physical Theory*, Athenium, 1962.

Brian Easlea, *Liberation and the Aims of Science: An Essay on Obstacles to the Building of a Beautiful World*, Chatto and Windus, 1973.

Arthur Eddington, *The Nature of the Physical World*, Dent, 1935.

Albert Einstein, *Ideas and Opinions*, Souvenir Press, 1973.

Albert Einstein, "Autobiographical Notes", in P.A. Schlipp (ed.) *Albert Einstein: Philosopher-Scientist*, Open Court, 1970, pp 3-94.

Albert Einstein and Leopold Infeld, *The Evolution of Physics: The Growth of Ideas From the Early Concepts of Relativity and Quanta*, Cambridge University Press, 1938.

Paul A. Feyerabend, "Problems of Empiricism", in Robert G. Colodny (ed.) *Beyond the Edge of Certainty*, Prentice-Hall, 1965, pp. 145-260.

Richard Feynman, *The Character of Physical Law*, M.I.T. Press, 1967.

Eva Figes, *Patriarchal Attitudes*, Panther Books Ltd., 1972.

Erich Fromm, *The Fear of Freedom*, Routledge and Kegan Paul, 1960.

Erich Fromm, *The Sane Society*, Routledge and Kegan Paul, 1963.

J. C. Graves, *The Conceptual Foundations of Contemporary Relativity Theory*, M.I.T. Press, 1971.

Daniel T. Gillespie, *A Quantum Mechanical Primer*, International Textbook Company Ltd., 1973.

David Hume, *A Treatise of Human Nature*, Dent, 1959.

Alistair Hardy, *The Living Stream: A Restatement of Evolution Theory and Its Relation to the Spirit of Man*, Collins, 1965.

G.H. Hardy, *A Mathematician's Apology*, Cambridge University Press, 1973.

G.O. Jones, J. Rotblat, G. Whitrow, *Atoms and the Universe*, Penguin Books Ltd., 1973.

R. Jungk, *Brighter than a Thousand Suns*, Penguin Books Ltd., 1970.

Edna E. Kramer, *The Nature and Growth of Modern Mathematics*, Hawthorn, 1970.

Arthur Koestler, *The Sleepwalkers*, Penguin Books Ltd., 1964.

Thomas S. Kuhn, *The Structure of Scientific Revolutions*, University of Chicago Press, 1970.

R.D. Laing, *The Divided Self*, Penguin Books Ltd., 1965.

Imre Lakatos, *Proofs and Refutations: The Logic of Mathematical Discovery*, Cambridge University Press, 1976.

N. Maxwell, 'Physics and Common Sense', B*ritish Journal for the Philosophy of Science* 16, 1966, pp. 295-311.

N. Maxwell, 'Understanding Sensations', *Australasian Journal of Philosophy* 46, 1968, pp. 127-45.

N. Maxwell, 'Can there be Necessary Connections between Successive Events?', *British Journal for the Philosophy of Science* 19, 1968, pp. 1-25 [Reprinted in Swinburne (1974), pp. 149-74].

N. Maxwell, 'A Critique of Popper's Views on Scientific Method', *Philosophy of Science* 39, 1972, pp. 131-52.

N. Maxwell, 'A New Look at the Quantum Mechanical Problem of Measurement', *American Journal of Physics* 40, 1972, pp. 1431-5.

N. Maxwell, 'Towards a Micro Realistic Version of Quantum Mechanics', *Foundations of Physics* 6, pp. 275-92 and 661-76.

N. Maxwell, 'The Rationality of Scientific Discovery', *Philosophy of Science* 41, 1974, pp. 123-53 and 247-95.

John Stuart Mill, *Autobiography*, Oxford University Press, 1971.

Elaine Morgan, *The Descent of Woman*, Corgi Books, 1973.

Robert M. Pirsig, *Zen and the Art of Motorcycle Maintenance: An Inquiry into Values*, The Bodley Head, 1974.

H. Poincaré, *Science and Hypothesis*, Dover Publications Inc., 1952.

Michael Polanyi, *Personal Knowledge: Towards a Post-Critical Philosophy*, Routledge and Kegan Paul, 1973.

G. Polya, *How to Solve It*, Anchor Books, 1945.

K.R. Popper, *The Open Society and Its Enemies*, Routledge and Kegan Paul, 1962.

K.R. Popper, *Conjectures and Refutations*, Routledge and Kegan Paul, 1963.

K.R. Popper, *Objective Knowledge: An Evolutionary Approach*, Oxford University Press, 1972.

K.R. Popper, *Unended Quest: An Intellectual Autobiography*, Fontana, 1976.

John Cowper Powys, *In Spite Of: A Philosophy for Everyman*, Village Press, 1974.

Oliver W. Sacks, *Awakenings*, Penguin Books Ltd., 1976.

E.F. Schumacher, *Small is Beautiful: A Study of Economics as if People Mattered,* Blond and Briggs, 1973.

J. Andrade e Silva and G. Lochak, *Quanta*, World University Library, 1969.

Isaac Bashevis Singer, *The Manor* and *The Estate*, Penguin Books Ltd., 1975.

R. Swinburne, (ed.) *The Justification of Induction*, Oxford University Press, 1974.

Thomas Szasz, *The Myth of Mental Illness: Foundations of a Theory of Personal Conduct*, Paladin, 1972.

Colin Turnbull, *The Forest People*, Picador, 1976.

Gerald Wick, *Elementary Particles: Frontiers of High Energy Physics*, Geoffrey Chapman, 1972.

Eugene Wigner, *Symmetries and Reflections: Scientific Essays*, M.I.T. Press, 1972.

Yevgeny Zamyatin, *We*, Penguin Books Ltd., 1972.

FURTHER READING POST 1976

Adair, R. (1987) *The Great Design*, Oxford University Press, Oxford.

Anderson, M. (1992) *Imposters in the Temple: American Intellectuals Are Destroying Our Universities and Cheating Our Students of Their Future*, Simon and Schuster, New York.

Attenborough, D. (1981) *Life on Earth*, Fontana, London.

Baars, B. J. (1988), *A Cognitive Theory of Consciousness*, Cambridge University Press, Cambridge.

Benton, M. (2003) *When Life Nearly Died*, Thames and Hudson, London.

Berlin, I. (1979) *Against the Current*, Hogarth Press, London.

Berlin, I. (1980) *Concepts and Categories*, Oxford University Press, Oxford.

Bond, E. J. (1983) *Reason and Value*, Cambridge University Press, Cambridge.

Brink, D. (1989) *Moral Realism and the Foundations of Ethics*, Cambridge University Press, Cambridge.

Brockway, G. (1995) *The End of Economic Man*, W. W. Norton, New York.

Burrows, B. et al. (1991) *Into the 21st Century*, Adamantine Press, London.

Carson, R. (1972) *Silent Spring*, Penguin, Harmondsworth (first published 1962).

Carter, R. (1998) *Mapping the Mind*, Weidenfeld and Nicolson, London.

Chisholm, J. S. (1999) *Death, Hope and Sex*, Cambridge University Press, Cambridge.

Collingridge, D. (1981) *The Social Control of Technology*, Open University Press, Milton Keynes.

The Committee.... (1981) *Hiroshima and Nagasaki: the Physical, Medical, and Social Effects of the Atomic Bombings*, Hutchinson, London.

Dawkins, R. (1978) *The Selfish Gene*, Paladin, London.

Deane-Drummond, C. (2006) *Wonder and Wisdom*, Novalis, Toronto.

Dennett, D. C. (2003) *Freedom Evolves*, Allen Lane, London.

Dixon, N. (1988) *Our Own Worst Enemy*, Futura, London.

Dubos, R. and Ward, B. (1972) *Only One Earth*, Penguin, Harmondsworth.

Dukas, H. and Hoffmann, B. (1979) *Albert Einstein: the Human Side*, Princeton University Press, Princeton.

Elms, D.G. (1989) 'Wisdom Engineering - The Methodology of Versatility', *Int. J. Appl. Engng. Ed.* Vol. 5, pp. 711-7.

Feyerabend, P. (1978) *Science in a Free Society*, New Left Books, London.

Gaa, J. et al. (1977) 'Value issues in science, technology and medicine', *Philosophy of Science*, 44, pp. 511-618.

Gascardi, A. (1999) *Consequences of Enlightenment*, Cambridge University Press, Cambridge.

Goodwin, P. (1982) *Nuclear War: the Facts*, Macmillan, London.

Gould, S. J. (1989) *Wonderful Life*, Hutchinson Radius, London.

Greene, B. (1999) *The Elegant Universe*, W. W. Norton, New York.

Gross, P. and N. Levitt (1994) *Higher Superstition: The Academic Left and Its Quarrels with Science*, John Hopkins University Press, Baltimore.

Gross, P., N. Levitt and M. Lewis (eds.) (1996) *The Flight from Science and Reason*, John Hopkins University Press, Baltimore.

Guth, A. (1997) *The Inflationary Universe*, Jonathan Cape, London.

Harris, M. (1979) *Cultural Materialism*, Random House, New York.

Harris, N. (1968) *Beliefs in Society*, Penguin, Harmondsworth.

Harrison, P. (1979) *Inside the Third World*, Penguin, Harmondsworth.

Higgins, R. (1978) *The Seventh Enemy: the Human Factor in the Global Crisis*, Hodder and Stoughton, London.

Howson, C. (2002) *Hume's Problem*, Oxford University Press, Oxford.

Kane, R. (1988) *The Significance of Free Will*, Oxford University Press, Oxford.

Kekes, J. (1985) 'The Fate of the Enlightenment Program', *Inquiry* 28, pp. 388-98.

Koertge, N. (1989) 'Review of From Knowledge to Wisdom', *Isis*

80, pp. 146-7.

Koertge, N. (ed.) (1998) *A House Built on Sand: Exposing Postmodernist Myths about Science*, Oxford University Press, Oxford.

Kuhn, T.S. (1977) *The Essential Tension*, Chicago University Press, Chicago.

Langley, C. (2005) *Soldiers in the Laboratory*, Scientists for Global Responsibility, Folkstone.

Little, M. (1994) 'Moral Realism', *Philosophical Books* 35, pp. 145-53 and 225-33.

Longuet-Higgins, C. (1984) 'For goodness sake', *Nature* 312, p. 204.

Maxwell, N. (1977) 'Articulating the Aims of Science', *Nature*, 265, p. 2.

Maxwell, N. (1980) 'Science, Reason, Knowledge and Wisdom: a Critique of Specialism', *Inquiry*, 23, pp. 19-81.

Maxwell, N. (1982) 'Instead of Particles and Fields: a Micro-Realistic Quantum "Smearon" Theory', *Foundations of Physics*, 12, pp. 607-31.

Maxwell, N. (1984) *From Knowledge to Wisdom: A Revolution in the Aims and Methods of Science*, Basil Blackwell, Oxford.

Maxwell, N. (1984) 'From Knowledge to Wisdom: Guiding Choices in Scientific Research', *Bulletin of Science, Technology and Society* 4, pp. 316-334.

Maxwell, N. (1985) 'Methodological Problems of Neuroscience', in D. Rose and V.G. Dobson (eds.) *Models of the Visual Cortex* John Wiley, Chichester, pp. 11-21.

Maxwell, N. (1985) 'Are Probabilism and Special Relativity Incompatible?', *Philosophy of Science* 52, pp. 23-44.

Maxwell, N. (1986) 'The Fate of the Enlightenment: Reply to Kekes', *Inquiry* 29, pp. 79-82.

Maxwell, N. 'Wanted: a new way of thinking', *New Scientist*, 14 May 1987, p. 63.

Maxwell, N. (1988) 'Quantum Propensiton Theory: A Testable Resolution of the Wave/Particle Dilemma', *British Journal for the Philosophy of Science* 39, pp. 1-50.

Maxwell, N. (1990) 'How Can We Build a Better World?' in J. Mittelstrass (ed.) *Einheit der Wissenschaften: Internationales*

Kolloquium der Akademie der Wissenschaften zu Berlin, Walter de Gruyter, Berlin and New York, pp. 388-427.

Maxwell, N. (1992) 'What Kind of Inquiry Can Best Help Us Create a Good World?', *Science, Technology and Human Values* 17, pp. 205-27.

Maxwell, N. (1992) 'What the Task of Creating Civilization has to Learn from the Success of Modern Science: Towards a New Enlightenment', *Reflections on Higher Education* 4, pp. 47-69.

Maxwell, N. (1993) 'Beyond Fapp: Three Approaches to Improving Orthodox Quantum Theory and An Experimental Test', in A. van der Merwe, F. Selleri and G. Tarozzi (eds.) *Bell's Theorem and the Foundations of Modern Physics*, World Scientific, Singapore, pp. 362-370.

Maxwell, N. (1993) 'Does Orthodox Quantum Theory Undermine, or Support, Scientific Realism?', *The Philosophical Quarterly* 43, pp. 139-57.

Maxwell, N. (1993) 'Induction and Scientific Realism: Einstein versus van Fraassen', *British Journal for the Philosophy of Science* 44, pp. 61-79, 81-101 and 275-305.

Maxwell, N. (1994) 'Particle Creation as the Quantum Condition for Probabilistic Events to Occur', *Physics Letters A* 187, pp. 351-355.

Maxwell, N. (Spring 1997) 'Science and the environment: A new enlightenment', *Science and Public Affairs*, pp. 50-56.

Maxwell, N. (1998) *The Comprehensibility of the Universe: A New Conception of Science*, Oxford University Press, Oxford (pbk. 2003).

Maxwell, N. (1999) 'Has Science Established that the Universe is Comprehensible?', *Cogito* 13, pp. 139-145.

Maxwell, N. (1999) 'Are There Objective Values?', *The Dalhousie Review* 79/3, pp. 301-317.

Maxwell, N. (2000) 'Can Humanity Learn to become Civilized? The Crisis of Science without Civilization', *Journal of Applied Philosophy*, 17, pp. 29-44.

Maxwell, N. (2000) 'A new conception of science, *Physics World* 13, No. 8, pp. 17-18.

Maxwell, N. (2000) 'The Mind-Body Problem and Explanatory Dualism', *Philosophy* 75, pp. 49-71.

Maxwell, N. (2001) *The Human World in the Physical Universe: Consciousness, Free Will and Evolution*, Rowman and Littlefield, Lanham, Maryland.

Maxwell, N. (2001) 'Can Humanity Learn to Create a Better World? The Crisis of Science without Wisdom', in *The Moral Universe*, T. Bentley and D. Stedman Jones (eds.) *Demos Collection* 16 (2001), pp. 149-156.

Maxwell, N. (25 May 2001) 'Wisdom and curiosity? I remember them well', *The Times Higher Education Supplement*, no. 1,488, p. 14.

Maxwell, N. (2002) 'Karl Raimund Popper', in *British Philosophers, 1800-2000*, P. Dematteis, P. Fosl and L. McHenry, (eds.), Bruccoli Clark Layman, Columbia, pp. 176-194. See also http://philsci-archive.pitt.edu/archive/00001686/ .

Maxwell, N. (2002) 'The Need for a Revolution in the Philosophy of Science', *Journal for General Philosophy of Science* 33, pp. 381-408. See also http://philsci-archive.pitt.edu/archive/00002449/

Maxwell, N. (Spring 2002) 'Science and meaning: why physics can coexist with consciousness', *The Philosophers' Magazine* 18, pp. 15-16.

Maxwell, N. (March/April 2002) 'Cutting God in Half', *Philosophy Now*, pp. 22-25.

Maxwell, N. (2003) 'Science, Knowledge, Wisdom and the Public Good', *Scientists for Global Responsibility Newsletter* 26, pp. 7-9.

Maxwell, N. (2003) 'Do Philosophers Love Wisdom', *The Philosophers' Magazine*, Issue 22, 2nd quarter, pp. 22-24.

Maxwell, N. (2003) 'Art as Its Own Interpretation', in A. Ritivoi (ed.) *Interpretation and Its Objects: Studies in the Philosophy of Michael Krausz*, Rodopi, Amsterdam, pp. 269-283.

Maxwell, N. (2004) 'Scientific Metaphysics', http://philsci-archive.pitt.edu/archive/00001674/

Maxwell, N. (2004) *Is Science Neurotic?*, Imperial College Press, London.

Maxwell, N. (2004) 'Does Probabilism Solve the Great Quantum Mystery?', *Theoria* 19/3, no. 51, pp. 321-336.

Maxwell, N. (2004) 'In Defence of Seeking Wisdom', *Meta-*

philosophy 35, pp. 733-743.

Maxwell, N. (2004) 'Non-Empirical Requirements Scientific Theories Must Satisfy: Simplicity, Unification, Explanation, Beauty', http://philsci-archive.pitt.edu/archive/00001759/

Maxwell, N. (2004) 'Comprehensibility rather than Beauty', http://philsci-archive.pitt.edu/ archive/00001770/

Maxwell, N. (2005) 'A Revolution for Science and the Humanities: From Knowledge to Wisdom', *Dialogue and Universalism*, vol. XV, no. 1-2, pp. 29-57. See also: http://philsci-archive.pitt.edu /archive/00001874/ .

Maxwell, N. (2005) 'Popper, Kuhn, Lakatos and Aim-Oriented Empiricism', *Philosophia* 32, nos. 1-4, pp. 181-239.

Maxwell, N. (2005) 'Philosophy Seminars for Five-Year-Olds', *Learning for Democracy*, Vol. 1, No. 2, pp. 71-77. [reprinted in *Gifted Education International*, 22/2-3 (2007), pp. 122-127].

Maxwell, N. (2005) 'The Enlightenment, Popper and Einstein', http//philsci-archive.pitt.edu/ archive/00002215/ .

Maxwell, N. (2005) 'A Mug's Game? Solving the Problem of Induction with Metaphysical Presuppositions', http://philsci-archive.pitt.edu/archive/00002230/ .

Maxwell, N. (2005) 'Three Philosophical Problems about Consciousness and Their Possible Resolution', http://philsci-archive.pitt.edu/archive/00002238/ .

Maxwell, N. (2005) 'Science under Attack', *The Philosophers' Magazine*, Issue 31, 3rd Quarter, pp. 37-41.

Maxwell, N. (2006) 'Learning to Live a Life of Value', in J. Merchey (ed.) *Living a Life of Value* (Values of the Wise Press, pp. 383-395.

Maxwell, N. (2006) 'Practical Certainty and Cosmological Conjectures', in M. Rahenfeld, ed., *Gibt es sicheres Wissen?*, Leipziger Universitätsverlag, Leipzig, pp. 44-59. See also:- http://philsci-archive.pitt.edu/archive/00002259/ .

Maxwell, N. (2006) 'Special Relativity, Time, Probabilism and Ultimate Reality', in D. Dieks, (ed.), *The Ontology of Spacetime*, Elsevier, B. V., pp. 229-245.

Maxwell, N. (2006) 'The Enlightenment Programme and Karl Popper', in *Karl Popper: A Centenary Assessment. Volume 1: Life and Times, Values in a World of Facts*, I. Jarvie, K. Milford and D. Miller, (eds.) Ashgate, London, pp. 177-190.

Maxwell, N. (2007) *From Knowledge to Wisdom: A Revolution for Science and the Humanities*, Pentire Press, London, second edition, revised, new introduction and three new chapters.

Maxwell, N. (2007) 'From Knowledge to Wisdom: The Need for an Academic Revolution', *London Review of Education*, 5/2, pp. 97-115 [reprinted in Maxwell, N. and R. Barnett (2008), pp. 1-19].

Maxwell, N. (2007) 'Can the World Learn Wisdom?', *Solidarity, Sustainability, and Non-Violence*, 3/4, http://www.pelicanweb. org/solisustv03n04maxwell.html .

Maxwell, N. (2007) 'The Disastrous War against Terrorism: Violence versus Enlightenment', in A. W. Merkidze (ed.) *Terrorism Issues*, Nova Science Publishers, New York. See also: http://www.nick-maxwell.demon.co.uk/Terrorism.htm .

Maxwell, N. and R. Barnett, (eds.) (2008) *Wisdom in the University*, Routledge, London.

Maxwell, N., (2008) 'Are Philosophers Responsible for Global Warming?', *Philosophy Now*, issue 65, pp. 12-13.

Maxwell, N. (2008) 'Do We Need a Scientific Revolution?', *Journal for Biological Physics and Chemistry*, 8/3, pp. 95-105.

Maxwell, N. (2008) Contribution to *How to Think About Science*, Ideas Transcripts, David Cayley, (ed.), Canadian Broadcasting Corporation, Toronto, 2008, pp. 212-220 (text of broadcast on 18 June 2008).

Maxwell, N. (2009) 'How Can Life of Value Best Flourish in the Real World?', in McHenry, L. (ed.) (2009), pp. 1-56.

Maxwell, N. (2009) 'Replies and Reflections', in McHenry, L. (2009), pp. 249-313.

Maxwell, N. (2009) 'Is the Quantum World Composed of Propensitons?', in M. Suárez (ed.) *Probabilities, Causes and Propensities in Physics*, Synthese Library, Boston.

Maxwell, N. (2009) 'Popper's Paradoxical Pursuit of Natural Philosophy', in *Cambridge Companion to Popper*, J. Shearmur and G. Stokes, (eds.) Cambridge University Press, Cambridge.

Maxwell, N. (2009) 'A Priori Conjectural Knowledge in Physics', in *What Place for the A Priori?*, M. Shaffer and M. Veber, (eds.) Open Court, La Salle, Illinois.

272

May, Robert (2005) *Threats to Tomorrow's World*, The Royal Society, London.

McAllister, J. (1996) *Beauty and Revolution in Science*, Cornell University Press, Ithaca.

McHenry, L. (ed.) (2009) *Science and the Pursuit of Wisdom: Studies in the Philosophy of Nicholas Maxwell*, Ontos Verlag, Frankfurt.

Midgley, M. (1986) 'Is wisdom forgotten?', *Culture, Education and Society* 40, pp. 425-7.

Midgley, M. (1989) *Wisdom, Information and Wonder*, Routledge, London.

Monbiot, G. (2006) *Heat*, Allen Lane, London.

Musgrave, A. (1993) *Common Sense, Science and Scepticism*, Cambridge University Press, Cambridge.

Norman, C. (1981) *The God that Limps*, Norton, New York.

O'Hear, A., (ed.) (2003) *Popper: Critical Assessments of Leading Philosophers*, Vol. II, Routledge, London.

Pais, A. (1980) 'Einstein on particles, fields, and the quantum theory', in H. Woolf (ed.), *Some Strangeness in the Proportion*, Addison-Wesley, Reading, Mass., pp. 197-251.

Pais, A. (1982) *Subtle is the Lord . . .* , Clarendon Press, Oxford.

Pearce, D. et al. (1989) *Blueprint for a Green Economy*, Earthscan, London.

Pearce, D. et al. eds. (1991) *Blueprint 2*, Earthscan, London.

Penrose, R. (2004) *The Road to Reality*, Jonathan Cape, London.

Popper, K.R. (1998) *The World of Parmenides*, Routledge, London.

Popper K.R. and J. Eccles (1977) *The Self and Its Brain*, Springer International, London.

Porritt, J. (1984) *Seeing Green*, Blackwell, Oxford.

Posner, M. (1993) 'Seeing the Mind', *Science* 262, pp. 673-4.

Rees, M. (2003), *Our Final Century*, Arrow Books, London.

Roszak, T. (1970) *The Making of a Counter Culture*, Faber, London.

Rotblat, J. (1983) *Scientists, the Arms Race and Disarmament*, Taylor and Francis, London.

Salmon, W. (1989) *Four Decades of Scientific Explanation*, University of Minnesota Press, Minneapolis.

Schilpp. P.A. (ed.) (1970) *Albert Einstein: Philosopher-Scientist*, Open Court, La Salle, Illinois.

Sen, A. (1987) *On Ethics and Economics*, Blackwell, Oxford.

Silk, J. (1980) *The Big Bang*, Freeman, San Francisco.

Smith, D. (2003) *The Atlas of War and Peace*, Earthscan, London.

Sokal, A. and J. Bricmont (1998) *Intellectual Impostures*, Profile Books, London.

Sternberg, R. (ed.) (1990) *Wisdom: Its Nature, Origins and Development*, Cambridge University Press, Cambridge.

Stiglitz, J. (2002) *Globalization and Its Discontents*, Penguin, London.

Stone, I. F. (1989) *The Trial of Socrates*, Picador, London.

Thorne, K. (1994) *Black Holes and Time Warps*, Picador, London.

Tyndall Centre, ed. (2006) *Truly Useful*, Tyndall Centre, UK.

Veneman, A. (2006) 'Achieving Millennium Goal 4', *The Lancet: Child Survival*, September 2006, pp. 4-7.

Ward, B. (1979) *Progress for a Small Planet*, Penguin, Harmondsworth.

Watkins, J. W. N. (1984) *Science and Scepticism*, Princeton University Press, Princeton.

Weart, S. (2003) *The Discovery of Global Warming*, Harvard University Press, Cambridge, Mass.

Weinberg, S. (1977) *The First Three Minutes*, Andre Deutsch, London.

Weinberg, S. (1993) *Dreams of a Final Theory*, Hutchinson, London.

Wilsdon, J. and R. Willis (2004) *See-through Science*, Demos, London.

Wolpert, L. (1993) *The Unnatural Nature of Science*, Faber and Faber, London.

World Health Report (2000), World Health Organization.

Yates, S., (1989) 'From Knowledge to Wisdom: Notes on Maxwell's Call for Intellectual Revolution', *Metaphilosophy* 20, pp. 371-86.

COMMENTS ON WORK BY

NICHOLAS MAXWELL

From Knowledge to Wisdom (1984; 2nd ed., 2007)

"a strong effort is needed if one is to stand back and clearly state the objections to the whole enormous tangle of misconceptions which surround the notion of science to-day. Maxwell has made that effort in this powerful, profound and important book."
Dr. Mary Midgley, *University Quarterly*

"The essential idea is really so simple, so transparently right ... It is a profound book, refreshingly unpretentious, and deserves to be read, refined and implemented."
Dr. Stewart Richards, *Annals of Science*

"Maxwell's book is a major contribution to current work on the intellectual status and social functions of science ... [It] comes as an enormous breath of fresh air, for here is a philosopher of science with enough backbone to offer root and branch criticism of scientific practices and to call for their reform."
Dr. David Collingridge, *Social Studies of Science*

"Maxwell has, I believe, written a very important book which will resonate in the years to come. For those who are not inextricably and cynically locked into the power and career structure of academia with its government-industrial-military connections, this is a book to read, think about, and act on."
Dr. Brian Easlea, *Journal of Applied Philosophy*

"This book is the work of an unashamed idealist; but it is none the worse for that. The author is a philosopher of science who holds the plain man's view that philosophy should be a guide to life, not just a cure for intellectual headaches. He believes, and argues with passion and conviction, that the abysmal failure of science to free society from poverty, hunger and fear is due to a fatal flaw in the

accepted aim of scientific endeavour – the acquisition and extension of knowledge. It is impossible to do Maxwell's argument justice in a few sentences, but, essentially, it is this. At the present time the pursuit of science – indeed the whole of academic inquiry – is largely dominated by 'the philosophy of knowledge'. At the heart of this philosophy is the assumption that knowledge is to be pursued for its own sake. But the pursuit of objective truth must not be distorted by human wishes and desires, so scientific research becomes divorced from human needs, and a well-intentioned impartiality gives way to a deplorable indifference to the human condition. The only escape is to reformulate the goals of science within a 'philosophy of wisdom', which puts human life first and gives 'absolute priority to the intellectual tasks of articulating our problems of living, proposing and criticizing possible solutions, possible and actual human actions'. The philosophy of wisdom commends itself, furthermore, not only to the heart but to the head: it gives science and scholarship a proper place in the human social order. . . Nicholas Maxwell has breached the conventions of philosophical writing by using, with intent, such loaded words as 'wisdom', 'suffering' and 'love'. 'That which is of value in existence, associated with human life, is inconceivably, unimaginably, richly diverse in character.' What an un-academic proposition to flow from the pen of a lecturer in the philosophy of science; but what a condemnation of the academic outlook, that this should be so. Mr. Maxwell is advocating nothing less than a revolution (based on reason, not on religious or Marxist doctrine) in our intellectual goals and methods of inquiry ... There are altogether too many symptoms of malaise in our science-based society for Nicholas Maxwell's diagnosis to be ignored."
Professor Christopher Longuet-Higgins, *Nature*

"Wisdom, as Maxwell's own experience shows, has been outlawed from the western academic and intellectual system ... In such a climate, Maxwell's effort to get a hearing on behalf of wisdom is indeed praiseworthy."
Dr. Ziauddin Sardar, *Inquiry*

"This book is a provocative and sustained argument for a 'revolution', a call for a 'sweeping, holistic change in the overall aims and methods of institutionalized inquiry and education, from knowledge to wisdom' ... Maxwell offers solid and convincing arguments for the exciting and important thesis that rational research and debate among professionals concerning values and their realization is both possible and ought to be undertaken."
Professor Jeff Foss, *Canadian Philosophical Review*

"Maxwell's argument ... is a powerful one. His critique of the underlying empiricism of the philosophy of knowledge is coherent and well argued, as is his defence of the philosophy of wisdom. Most interesting, perhaps, from a philosophical viewpoint, is his analysis of the social and human sciences and the humanities, which have always posed problems to more orthodox philosophers, wishing to reconcile them with the natural sciences. In Maxwell's schema they pose no such problems, featuring primarily ... as methodologies, aiding our pursuit of our diverse social and personal endeavours. This is an exciting and important work, which should be read by all students of the philosophy of science. It also provides a framework for historical analysis and should be of interest to all but the most blinkered of historians of science and philosophy."
Dr. John Hendry, *British Journal for the History of Science*

"... a major source of priorities, funds and graduates' jobs in 'pure science' is military ... this aspect of science is deemed irrelevant by the overwhelming majority of those who research, teach, sociologize, philosophise or moralize about science. What are we to make of such a phenomenon? It is in part a political situation, in its causes and effects; but it is also philosophical, and this is Nick Maxwell's point of focus. Such a gigantic co-operative endeavour of concealment, amounting to a huge deception, could be accomplished naturally by all educated, humane participants, a 'conspiracy needing no conspirators', only because their 'philosophy of knowledge' envelops them in the assurance that their directors, paymasters and employers have nothing to do with

the real thing – the research. This, to me, is the heart of Maxwell's message."
Dr. Jerry Ravetz, *British Journal for the Philosophy of Science*

"This book is written in simple straightforward language ... The style is passionate, committed, serious; it communicates Maxwell's conviction that we are in deep trouble, that there is a remedy available, and that it is ingrained bad intellectual habits that prevent us from improving our lot ... Maxwell is raising an important and fundamental question and things are not going so well for us that we should afford the luxury of listening only to well-tempered answers."
Professor John Kekes, *Inquiry*

"Because Maxwell so obviously understands and loves science as practiced, say, by an Einstein, his criticisms of current science seem to arise out of a sadness at missed opportunities rather than hostility ... I found Maxwell's exposition and critique of the current state of establishment science to be clear and convincing ... Maxwell is right to remind us that in an age of Star Wars and impending ecological disaster, talk of the positive potential of means-oriented science can easily become an escapist fantasy."
Professor Noretta Koertge, *Isis*

"In an admirable book called *From Knowledge to Wisdom*, Nicholas Maxwell has argued that the radical, wasteful misdirection of our whole academic effort is actually a central cause of the sorrows and dangers of our age . . .Thinking out how to live is a more basic and urgent use of the human intellect than the discovery of any fact whatsoever, and the considerations it reveals ought to guide us in our search for knowledge. . . In arguing this point . . . Maxwell proposes that we should replace the notion of aiming at knowledge by that of aiming at wisdom. I think this is basically the right proposal. . . Maxwell is surely right in saying that [the distorted pursuit of knowledge], because it wastes our intellectual powers, has played a serious part in distorting our lives."
Mary Midgley, *Wisdom, Information and Wonder*

"[T]here is...much of interest and, yes, much of value in this book...Maxwell is one of those rare professional philosophers who sees a problem in the divorce between thought and life which has characterized much of modern philosophy (and on both sides of the English channel, not merely in the so-called 'analytic' tradition'); he wishes to see thought applied to life and used to improve it. As a result, many of the issues he raises are of the first importance ... He has . . produced a work which should give all philosophers and philosophically-minded scientists cause for reflection on their various endeavors; in particular, it should give philosophers who are content to be specialists a few sleepless nights."
Professor Steven Yates, *Metaphilosophy*

"Maxwell [argues for] an "intellectual revolution" that will affect the fundamental methods of inquiry of science, technology, scholarship and education, looking not for knowledge for knowledge's sake, but for wisdom, which he says is more rational and of greater human value and holds the potential to alleviate human problems and institute social change. A humanist and philosopher, Maxwell presents his ideas with eloquence and conviction. This book will appeal to persons in many different disciplines – from science to social studies."
American Library Association

"Maxwell's thesis is that the evident failure of science to free society from poverty, hunger and the threat of extinction results from a 'fatal flaw in the accepted aim of scientific endeavour'. . . It is precisely because of 'the accepted aim' that acquisition of knowledge, which presumably originated as an essential strategy for survival, has given rise to the relentless pursuit of new and better ways of achieving the exact opposite. . . For Maxwell, the solution is obvious – a radically new approach to the whole business of intellectual inquiry. . . It is hard to argue with these aims . . . If we could only change the way people feel, Maxwell's solution would be easier, if not easy."
Professor Norman F. Dixon, *Our Own Worst Enemy*

"a sustained piece of philosophical reasoning which makes a real contribution to the reinstatement of philosophy as a central concern. We need to follow Maxwell's lead in constructing a philosophy of wisdom."
P. Eichman, *Perspectives on Science and Christian Faith*

"Any philosopher or other person who seeks wisdom should read this book. Any educator who loves education – especially those in leadership positions – should read this book. Anyone who wants to understand an important source of modern human malaise should read this book. And anyone trying to figure out why, in a world that produces so many technical wonders, there is such an immense "wisdom gap" should read this book. In *From Knowledge to Wisdom: A Revolution for Science and the Humanities*, Second Edition ...Nicholas Maxwell presents a compelling, wise, humane, and timely argument for a shift in our fundamental "aim of inquiry" from that of knowledge to that of wisdom."
Jeff Huggins *Metapsychology*

"In this book, Nicholas Maxwell argues powerfully for an intellectual "revolution" transforming all branches of science and technology. Unlike such revolutions as those described by Thomas Kuhn, which affect knowledge about some aspect of the physical world, Maxwell's revolution involves radical changes in the aims, methods, and products of scientific inquiry, changes that will give priority to the personal and social problems that people face in their efforts to achieve what is valuable and desirable."
Professor George Kneller, *Canadian Journal of Education*

"Nicholas Maxwell (1984) defines freedom as 'the capacity to achieve what is of value in a range of circumstances'. I think this is about as good a short definition of freedom as could be. In particular, it appropriately leaves wide open the question of just what is of value. Our unique ability to reconsider our deepest convictions about what makes life worth living obliges us to take seriously the discovery that there is no palpable constraint on what we can consider."
Professor Daniel Dennett, *Freedom Evolving*

"The Rationality of Scientific Discovery", *Philosophy of Science* (1974)

"Maxwell's theory of aim-oriented empiricism is the outstanding work on scientific change since Lakatos, and his thesis is surely correct. Scientific growth should be rationally directed through the discussion, choice, and modification of aim-incorporating blueprints rather than left to haphazard competition among research traditions seeking empirical success alone. . . Of the theories of scientific change and rationality that I know, Maxwell's is my first choice. It is broad in scope, closely and powerfully argued, and is in keeping with the purpose of this book, which is to see science in its totality. No other theory provides, as Maxwell's does in principle, for the rational direction of the overall growth of science."
Professor George F. Kneller, *Science as a Human Endeavor*

"As Nicholas Maxwell has suggested, if we make one crucial assumption about the purpose of science, then the possibility arises that some paradigms and theories can be evaluated even prior to the examination of their substantive products. This one crucial assumption is that the overall aim of science is to discover the maximum amount of order inherent in the universe or in any field of inquiry. Maxwell calls this 'aim-oriented empiricism'. . . I agree with Maxwell's evaluation of the importance of coherent aim-oriented paradigms as a criterion of science. . . The time is ripe, therefore, to replace the incoherent and unconscious paradigms under whose auspices most anthropologists conduct their research with explicit descriptions of basic objectives, rules, and assumptions. That is why I have written this book."
Professor Marvin Harris, *Cultural Materialism*

The Comprehensibility of the Universe: A New Conception of Science (1998)

"Nicholas Maxwell's ambitious aim is to reform not only our philosophical understanding of science but the methodology of scientists themselves ... Maxwell's aim-oriented empiricism [is]

281

intelligible and persuasive ... the main ideas are important and appealing ... an important contribution to the philosophy of physics."
J. J. C. Smart, *British Journal for the Philosophy of Science*

"Maxwell has clearly spent a lifetime thinking about these matters and passionately seeks a philosophical conception of science that will aid in the development of an intelligible physical worldview. He has much of interest to say about the development of physical thought since Newton. His comprehensive coverage and sophisticated treatment of basic problems within the philosophy of science make the book well worth studying for philosophers of science as well as for scientists interested in philosophical and methodological matters pertaining to science."
Professor Cory F. Juhl, *International Philosophical Quarterly*

"Maxwell performs a heroic feat in making the physics accessible to the non-physicist ... Philosophically, there is much here to stimulate and provoke . . . there are rewarding comparisons to be made between the functional roles assigned to Maxwell's metaphysical "blueprints" and Thomas Kuhn's paradigms, as well as between Maxwell's description of theoretical development and Imre Lakatos's methodology of scientific research programmes."
Dr.Anjan Chakravartty, *Times Higher Education Supplement*

"some of [Maxwell's] insights are of everlasting importance to the philosophy of science, the fact that he stands on the shoulders of giants (Hume, Popper) notwithstanding . . . My overall conclusion is that Universe is an ideal book for a reading group in philosophy of science or in philosophy of physics. Many of the pressing problems of the philosophy of science are discussed in a lively manner, controversial solutions are passionately defended and some new insights are provided; in particular the chapter on simplicity in physics deserves to be read by all philosophers of physics."
Dr. F. A. Muller, *Studies in History and Philosophy of Modern Physics*

"Maxwell ... has shown that it is absurd to believe that science can proceed without some basic assumptions about the comprehensibility of the universe . . . Throughout this book, Maxwell has meticulously argued for the superiority of his view by providing detailed examples from the history of physics and mathematics . . . The Comprehensibility of the Universe attempts to resurrect an ideal of modern philosophy: to make rational sense of science by offering a philosophical program for improving our knowledge and understanding of the universe. It is a consistent plea for articulating the metaphysical presuppositions of modern science and offers a cure for the theoretical schizophrenia resulting from acceptance of incoherent principles at the base of scientific theory."
Professor Leemon McHenry, *Mind*

"This admirably ambitious book contains more thought-provoking material than can even be mentioned here. Maxwell's treatment of the descriptive problem of simplicity, and his novel proposals about quantum mechanics deserve special note. In his view the simplicity of a theory is (and should be) judged by the degree to which it exemplifies the current blueprint of physicalism, that blueprint determining the terminology in which the theory and its rivals should be compared. This means that the simplicity of a theory amounts to the unity of its ontology, a view that allows Maxwell to offer an explanation of our conflicting intuitions that terminology matters to simplicity, and that it is utterly irrelevant. Maxwell's distinctive views about what is wrong with quantum mechanics grow out of his adherence to aim-oriented empiricism: the much-discussed problem of measurement is for him a superficial consequence of the deeper problem that the ontology of the theory is not unified, in that no one understands how one entity could be both a wave and a particle. In response to this problem Maxwell finds between the metaphysical cracks a way to fuse micro-realism and probabilism, which leads him to a proposal to solve the measurement problem by supplementing quantum mechanics with a collapse theory distinct from the recent and popular one of Ghirardi, Rimini and Weber. Maxwell's highly informed discussions of the changing ontologies of various modern

physical theories are enjoyable, and the physical and mathematical appendix of the book should be a great help to the beginner."
Professor Sherrilyn Roush, *The Philosophical Review*

"Nicholas Maxwell has struck an excellent balance between science and philosophy . . . The detailed discussions of theoretical unification in physics - from Newton, Maxwell and Einstein to Feynman, Weinberg and Salam - form some of the best material in the book. Maxwell is good at explaining physics . . . Through the interplay of metaphysical assumptions, at varying distances from the empirical evidence Maxwell shows, rather convincingly, that in the pursuit of rational science the inference from the evidence to a small number of acceptable theories, out of the pool of rival ones, is justifiable . . . Its greatest virtue is the detailed programme for a modern version of natural philosophy. Along the way, Maxwell homes in on the notion of comprehensibility by the exclusion of less attractive alternatives. In an age of excessive specialization the book offers a timely reminder of the close link between science and philosophy. There is a beautiful balance between concrete science and abstract philosophy . . . In the "excellently written Appendix some of the basic mathematical technicalities, including the principles of quantum mechanics, are very well explained . . . Einstein held that 'epistemology without science becomes an empty scheme' while 'science without epistemology is primitive and muddled'. Maxwell's new book is a long-running commentary on this aphorism."
Dr. Friedel Weinert, *Philosophy*

"In *The Comprehensibility of the Universe*, Nicholas Maxwell develops a bold, new conception of the relationship between philosophy and science...Maxwell has a metaphysically rich, evolutionary vision of the self-correcting nature of science...The work is important...An added benefit of Maxwell's analysis...is the possibility of a positive, fruitful relationship to emerge between science and the philosophy of science...his important and timely critique of the reigning empiricist orthodoxy...what does it mean to say simplicity is a theoretical virtue? And why should we prefer simple to complex theories? Maxwell provides an admirable discussion of these issues. He also provides a useful discussion of simplicity in the context of theory unification – simple theories are unifying theories – and illustrates his points with examples drawn from Newtonian physics and Maxwellian electrodynamics...It is hard to do justice to the richness of Maxwell's discussion in this chapter. I can only say that this is a chapter that will repay serious study...Maxwell turns his attention to issues surrounding the theoretical character of evidence, the idea of scientific progress and the question as to whether there is a method of discovery....The discussion of these matters – as with the other topics covered in this book – is conceptually rich and technically sophisticated. A useful antidote, in fact, to the settled orthodoxy surrounding these philosophical issues...Maxwell has written a book that aims to put the metaphysics back in physics. It is ambitious in scope, well-argued, and deserves to be seriously studied."

Professor Niall Shanks, *Metascience*

The Human World in the Physical Universe: Consciousness, Free Will and Evolution (2001)

"Ambitious and carefully-argued...I strongly recommend this book. It presents a version of compatibilism that attempts to do real justice to common sense ideas of free will, value, and meaning, and...it deals with many aspects of the most fundamental problems of existence."

Dr. David Hodgson, *Journal of Consciousness Studies*

"Maxwell has not only succeeded in bringing together the various different subjects that make up the human world/physical universe problem in a single volume, he has done so in a comprehensive, lucid and, above all, readable way."

Dr. M. Iredale, *Trends in Cognitive Sciences*

"...a bald summary of this interesting and passionately-argued book does insufficient justice to the subtlety of many of the detailed arguments it contains."

Professor Bernard Harrison, *Mind*

"Nicholas Maxwell takes on the ambitious project of explaining, both epistemologically and metaphysically, the physical universe and human existence within it. His vision is appealing; he unites the physical and the personal by means of the concepts of aim and value, which he sees as the keys to explaining traditional physical puzzles. Given the current popularity of theories of goal-oriented dynamical systems in biology and cognitive science, this approach is timely. . . The most admirable aspect of this book is the willingness to confront every important aspect of human existence in the physical universe, and the recognition that in a complete explanation, all these aspects must be covered. Maxwell lays out the whole field, and thus provides a valuable map of the problem space that any philosopher must understand in order to resolve it in whole or in part."
Professor Natika Newton, *Philosophical Psychology*

"This is a very complex and rich book. Maxwell convincingly explains why we should and how we can overcome the 'unnatural' segregation of science and philosophy that is the legacy of analytic philosophy. His critique of standard empiricism and defence of aim-oriented empiricism are especially stimulating"
Professor Thomas Bittner, *Philosophical Books*

"I recommend reading The Human World in the Physical Universe ... for a number of reasons. First, [it] ... provides the best entrance to Maxwell's world of thought. Secondly, [it] contains a succinct but certainly not too-detailed overview of the various problems and positions in the currently flourishing philosophy of mind. Thirdly, it shows that despite the fact that many philosophers have declared Cartesian Dualism dead time and again, with some adjustments, the Cartesian view remains powerful and can compete effortlessly with other extant views."
Dr. F. A. Muller, *Studies in History and Philosophy of Modern Physics*

"Some philosophers like neat arguments that address small questions comprehensively. Maxwell's book is not for them. The

Human World in the Physical Universe instead addresses big problems with broad brushstrokes."
Dr. Rachel Cooper, *Metascience*

"A solid work of original thinking."
Professor L. McHenry, *Choice*

Is Science Neurotic? (2004)

"This book is bursting with intellectual energy and ambition...[It] provides a good account of issues needing debate. In accessible language, Maxwell articulates many of today's key scientific and social issues...his methodical analysis of topics such as induction and unity, his historical perspective on the Enlightenment, his opinions on string theory and his identification of the most important problems of living are absorbing and insightful."
Clare McNiven, *Journal of Consciousness Studies*

"Is science neurotic? Yes, says Nicholas Maxwell, and the sooner we acknowledge it and understand the reasons why, the better it will be for academic inquiry generally and, indeed, for the whole of humankind. This is a bold claim ... But it is also realistic and deserves to be taken very seriously ... My summary in no way does justice to the strength and detail of Maxwell's well crafted arguments ... I found the book fascinating, stimulating and convincing ... after reading this book, I have come to see the profound importance of its central message."
Dr. Mathew Iredale, *The Philosopher's Magazine*

"... the title *Is Science Neurotic?* could be rewritten to read *Is Academe Neurotic?* since this book goes far beyond the science wars to condemn, in large, sweeping gestures, all of modern academic inquiry. The sweeping gestures are refreshing and exciting to read in the current climate of specialised, technical, philosophical writing. Stylistically, Maxwell writes like someone following Popper or Feyerabend, who understood the philosopher to be improving the World, rather than contributing to a small piece of one of many debates, each of which can be understood

only by the small number of its participants.... In spite of this, the argument is complex, graceful, and its finer points are quite subtle.... The book's final chapter calls for nothing less than revolution in academia, including the very meaning of academic life and work, as well as a list of the nine most serious problems facing the contemporary world - problems which it is the task of academia to articulate, analyse, and attempt to solve. This chapter sums up what the reader has felt all along: that this is not really a work of philosophy of science, but a work of 'Philosophy', which addresses 'Big Questions' and answers them without hesitation.... I enjoyed the book as a whole for its intelligence, courageous spirit, and refusal to participate in the specialisation and elitism of the current academic climate.... it is a book that can be enjoyed by any intelligent lay-reader. It is a good book to assign to students for these reasons, as well - it will get them thinking about questions like: What is science for? What is philosophy for? Why should we think? Why should we learn? How can academia contribute of the welfare of people? ... the feeling with which this book leaves the reader [is] that these are the questions in which philosophy is grounded and which it ought never to attempt to leave behind."
Margret Grebowicz, *Metascience*

"Maxwell's fundamental idea is so obvious that it has escaped notice. But acceptance of the idea requires nothing short of a complete revolution for the disciplines. Science should become more intellectually honest about its metaphysical presuppositions and its involvement in contributing to human value. Following this first step it cures itself of its irrational repressed aims and is empowered to progress to a more civilized world."
Professor Leemon McHenry, *Review of Metaphysics*

"Maxwell argues that the metaphysical assumptions underlying present-day scientific inquiry, referred to as standard empiricism or SE, have led to ominous irrationality. Hence the alarmingly provocative title; hence also-the argument carries this far-the sad state of the world today. Nor is Maxwell above invoking, as a parallel example to science's besetting "neurosis," the irrational behavior of Oedipus as Freud saw him: unintentionally yet intentionally slaying his father for love of his mother (Mother Earth?). Maxwell proposes replacing SE with his own metaphysical remedy, aim-oriented empiricism, or AOE. Since science does not acknowledge metaphysical presumptions and therefore disallows questioning them – they are, by definition, outside the realm of scientific investigation – Maxwell has experienced, over the 30-plus years of his professional life, scholarly rejection, which perhaps explains his occasional shrill tone. But he is a passionate and, despite everything, optimistic idealist. Maxwell claims that AOE, if adopted, will help deal with major survival problems such as global warming, Third World poverty, and nuclear disarmament, and science itself will become wisdom-oriented rather than knowledge-oriented – a good thing. A large appendix, about a third of the book, fleshes the argument out in technical, epistemological terms. Summing Up: Recommended. General readers; graduate students; faculty."

Professor M. Schiff, *Choice*

"*Is Science Neurotic?* … is a rare and refreshing text that convincingly argues for a new conception of scientific empiricism that demands a re-evaluation of what [science and philosophy] can contribute to one another and of what they, and all academia, can contribute to humanity… Is Science Neurotic? is primarily a philosophy of science text, but it is clear that Maxwell is also appealing to scientists. The clear and concise style of the text's four main chapters make them accessible to anyone even vaguely familiar with philosophical writing and physics… it is quite inspiring to read a sound critique of the fragmented state of academia and an appeal to academia to promote and contribute to social change."

Sarah Smellie, *Canadian Undergraduate Physics Journal*

"Maxwell's aspirations are extraordinarily and admirably ambitious. He intends to contribute towards articulating and bringing about a form of social progress that embodies rationality and wisdom... by raising the question of how to integrate science into wisdom-inquiry and constructing novel and challenging arguments in answer to it, Maxwell is drawing attention to issues

that need urgent attention in the philosophy of science."
Professor Hugh Lacey, *Mind*

"Maxwell has written a very important book … Maxwell eloquently discusses the astonishing advances and the terrifying realities of science without global wisdom. While science has brought forth significant advancements for society, it has also unleashed the potential for annihilation. Wisdom is now, as he puts it, not a luxury but a necessity … Maxwell's book is first-rate. It demonstrates his erudition and devotion to his ideal of developing wisdom in students. Maxwell expertly discusses basic problems in our intellectual goals and methods of inquiry."
Professor Joseph Davidow, *Learning for Democracy*

"My judgement of this book is favourable...[Maxwell's] heart is in the right place, as he casts a friendly but highly critical eye on the Enlightenment Movement. 'We suffer, not from too much scientific rationality, but from not enough' he says...recommending a massive cooperation between science and the humanities...The book's style is refreshingly simple, clear".
Joseph Agassi, *Philosophy of Science*

"Nicholas Maxwell's book passionately embraces Francis Bacon's dictum that '[t]he true and legitimate goal of the sciences is to endow human life with new discoveries and resources'. The book's scope is commendable. It offers a thorough critique of the contemporary philosophy and practice of both natural (Chapters 1 & 2) and social science (Chapter 3), and suggests a remedy for what the author believes is the neurotic repression of the aforementioned Baconian aims."
Slobadan Perovic, *British Journal for the Philosophy of Science*

www.ingramcontent.com/pod-product-compliance
Lightning Source LLC
Chambersburg PA
CBHW060228050426
42448CB00009B/1345